The Mechanical Properties of Materials

Methuen Studies in Science

This series provides students with concise, introductory surveys of important topics in the physical, chemical and biological sciences. The series is designed to assist sixth form students preparing for entry to university or college, and to meet the needs of university students preparing for more advanced studies.

Chemical Equilibrium	J. S. Coe
Enzymes	Alan D. B. Malcolm
Energy in Chemistry	B. E. Dawson
Alternating Currents	J. M. Gregory
Nature Conservation	W. M. M. Baron
Logical Control Systems	Keith Morphew
Aspects of Isomerism	Peter Uzzell
The Inorganic Chemistry of the Non-metals	John Emsley
Atomic and Molecular Weight Determination	R. B. Moyes
The Mechanical Properties of Materials	R. A. Farrar

Forthcoming

The States of Matter	A. Ralph Morgan
Oscillations	Ian B. Hopley
Transition Metal Chemistry	Jeff Thompson
Crystals and their Structures	I. F. Roberts
Catalysis in Chemistry	A. J. B. Robertson
Molecular Spectroscopy	Aline Bradshaw
Kinetics and Mechanisms of Reactions	B. E. Dawson
Waves	Ian B. Hopley
Radioisotopes	D. J. Hornsey
Fundamental Electrostatics	S. W. Hockey

Methuen Studies in Science

The Mechanical Properties of Materials

ROY A. FARRAR Ph.D., A.I.M.

Lecturer in the Department of Engineering Materials
Southampton University

Methuen Educational Ltd

LONDON · TORONTO · SYDNEY · WELLINGTON

First published 1971
by Methuen Educational Ltd
11 New Fetter Lane, London EC4

Printed in Great Britain by
William Clowes & Sons Ltd.,
London, Colchester and Beccles

SBN 423 83300 6 non net
423 86100 X net

Distributed in the U.S.A. by
Barnes and Noble Inc.

Contents

Preface

Although I am a metallurgist by training, I have, in this book, dealt unashamedly with the mechanical properties of all sorts of materials. I have done this because the déformation processes that occur at the microstructural and atomic levels and which control the macroscopic properties are common to all materials whether they are polymers, ceramics or metals.

I have sought to write the book at an introductory level that should appeal to both students taking the Nuffield A level Physical Sciences, and students at technical colleges and universities who are taking a preliminary course in the subject.

It has been my aim to show the reader the reason why a particular piece of material under stress behaves in the way it does and therefore I see the book in three main parts. The first two chapters introduce the subject and lay a foundation of simple atomic theory and crystal structures. The second section (chapters 3-5) deals with the particular materials – metals, polymers and ceramics – and shows how imperfections such as microcracks and dislocations control the properties we can measure in the laboratory.

Because materials are tangible things numerous experiments are suggested throughout the text to help the readers who have access to some simple equipment to test the concepts involved and to find out for themselves the fascinating things that occur when a material is stressed.

The final section is a brief resumé of some of the testing methods available and contains some suggestions as to the ways such variables as the temperature of testing may be investigated.

It is my hope that the next time the reader breaks something, he will at least know why it broke, even if he is not able to repair it.

I would like to acknowledge all the efforts and labours of my wife Sheila in the production of this book and also the many hours I spent with my colleague Dr D. A. Wigley discussing what really happens when a material is subjected to stress.

Southampton ROY A. FARRAR

1 The materials world

In the early days of man's civilization he used only the simple materials such as stone, bone and wood that were abundant around his dwelling place. With the discovery of native metals such as copper and gold he was able to fashion things for the first time in a ductile material and it is known that very early in Egypt the art of beating gold into very thin sheets was employed to provide not only an artistic but a corrosion-resistant finish to less durable materials. He learned to fire the raw clay and create a true ceramic which he could glaze with a low-melting-point glass and he learnt the art of smelting the ores of copper and iron and the production of alloys. Soon he discovered his ability to improve on his original materials by suitable heat-treatment and design, as the production of suits of armour and all sorts of weapons testify. With the advent of the industrial revolution came the production of vast tonnages of iron and steel; soon bridges and railways were being built, and every corner of the civilized world enjoyed the benefits of the materials that were easily fabricated into consumer articles. In the last few decades has come the discovery and exploitation of plastics and composite materials to add to the impressive list. It is only in the last sixty to seventy years however that we have begun to understand the properties of these materials and appreciate the connection between the variables such as chemical bonding, the melting point, crystalline imperfections, and strength.

Our modern world is a *materials* world. One only has to look around the average kitchen to see the wide range of materials that are being used. Steel in its many forms provides the refrigerator body, the window frames, screws and nails; stainless steel is used for sink units and cutlery. Copper of course is present in the electrical wiring in our many labour-saving devices. Plastic washing-up bowls, yogurt cups and containers are in abundance and new plastics such as polytetrafluorethylene (PTFE) are used to provide non-stick surfaces on cooking pots. Glasses and ceramics provide not only windows but the enamel coating on the cooker, the flame nozzle on the central-heating boiler and, in a different form, the bricks and concrete from which the house is built.

Our modern motor cars contain a bewildering array of materials, including many types of polymers and rubbers as well as the numerous metallic alloys in the engine and body. Aircraft which were traditionally constructed of aluminium alloys are now employing advanced stainless steels and titanium alloys and high-temperature plastics.

The range of moduli (which is related to the cohesive strength) exhibited by different materials is illustrated in Fig. 1, and it is obvious from this that there is a general correlation between the elastic modulus and the melting point and that

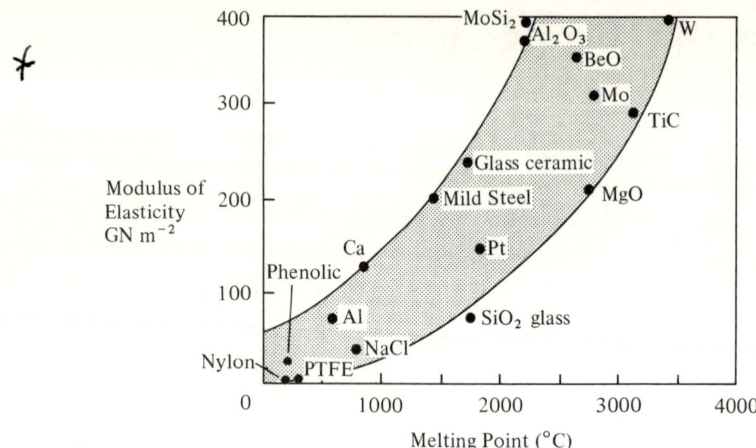

Fig. 1. Relationship between elastic modulus and melting point for different materials

those materials which possess the strongest chemical bonding exhibit both high moduli and melting points. It is interesting to notice that *ordinary* amorphous glass, which is based on the SiO_2 molecule, is a comparatively weak material and falls outside this general trend, whereas the new *ceramic glasses,* which are crystalline, are much stronger and compare favourably with materials such as mild steel.

In order to understand the range of mechanical properties exhibited by metals, polymers and ceramics, we will proceed in this monograph to examine, in Chapter 2, the nature of their chemical bonding and microstructure and then, in the later chapters, to see how these variables affect the specific mechanical properties of each class of materials.

2 Atomic bonding and structure of materials

Bonding

Molecular bonding and cohesion in solids occurs because the energy of the atoms is lower in the aggregate state than it is when the atoms are separated by large distances; as may be expected the interaction energy is at a minimum when the attractive and repulsive forces between adjacent atoms or molecules are equal and the slope of the energy distance curve is zero.

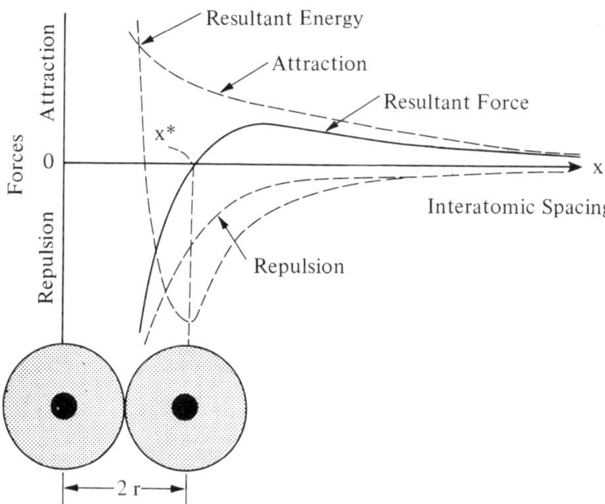

Fig. 2. Variation of potential energy and force with the distance between two atoms

In Fig. 2 typical curves are plotted which show how both the attractive and repulsive forces vary with the interatomic distance x; notice that the attractive force varies quite slowly with distance, but the repulsive force is increasing rapidly as x becomes small owing to the overlapping of the outer electrons. The equilibrium separation x^* is found where the attractive and repulsive forces are equal and the potential energy is at a minimum.

The various types of attractive bond may be classed as either strong *primary* bonds or weaker *secondary* bonds. The *primary* bonds include (a) **ionic** or electrovalent bonds between ions of opposite charge, (b) **covalent** or homopolar bonds between neutral atoms, and (c) **metallic** bonds between ions of the same charge. The secondary bonds are usually **van der Waals'**. All these bonds represent extremes and in a given material they may be present singly or in an intermediate form, or different bonds may act in different directions in the same aggregate.

PRIMARY BONDS The ionic bond is the electrostatic attraction of an electropositive with an electronegative ion. Fig. 3a indicates the transfer of the valency electron from the outer shell of an electropositive element such as sodium to the valency shell of electronegative chlorine, so producing Na^+ and Cl^- ions. This type of bond is found in inorganic salts such as the alkali halides and in ceramic materials such as MgO and Al_2O_3. The ionic bond is non-directional, and owing to the presence of the ions, ionically bonded materials can conduct electricity to a limited extent when in solution or at temperatures near to the melting point of the solid (typical resistivity – 10^3 to 10^5 ohm-m).

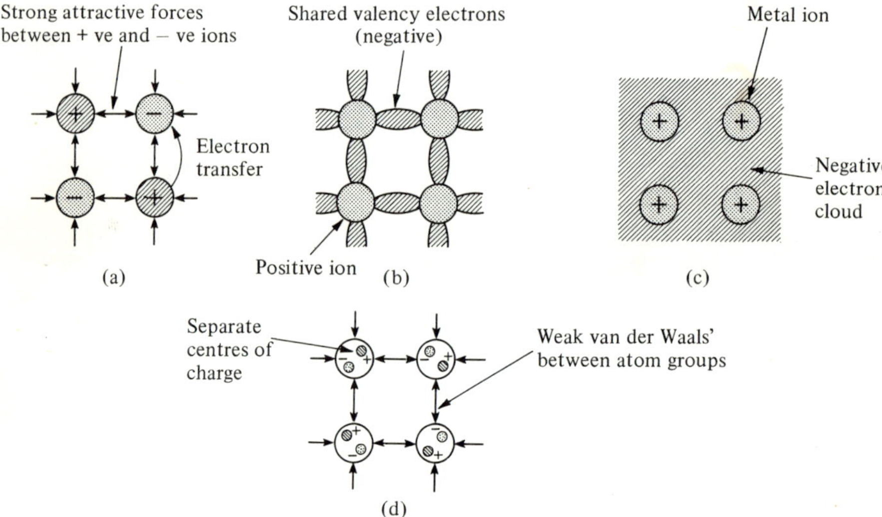

Fig. 3. The nature of the four types of bonding in solids: (a) ionic; (b) covalent; (c) metallic; and (d) van der Waals'

Another *primary* force of attraction is the covalent bond, which is formed when neighbouring atoms share their valency electrons in order to produce a completely full valency shell (in particular 8 electrons). These highly directional bonds (Fig. 3b) are formed between the two atoms in the hydrogen molecule, as well as in compounds such as CH_4 (methane), the attractive force in this case being due to the negatively charged shared electrons situated between the positively charged nuclei. This bond, which does not show electronic conduction (typical resistivity – 10^{17} to 10^{22} ohm-m), can provide strong attractive forces, as may be seen in the diamond form of carbon which possesses great hardness and is only thermally disrupted above 3300°C. However not all materials with covalent bonds have high melting and boiling points, or great strength. Methane for example has four covalent bonds between the carbon and hydrogen atoms, but the resulting molecule cannot form strong primary bonds with adjacent molecules and hence the intermolecular bonds, which are normally van der Waals' as shown in Fig. 3d, can be easily broken as is confirmed by the low boiling point – 161°C observed for liquid methane.

Metallic bonds are formed between elements whose valency electrons require low energies to remove them, to leave an assembly of positive ions and free electrons as illustrated in Fig. 3c. The ions being identical are free to move in the metallic lattice under the action of stress, so giving rise to the typical ductility of a metal, while the free electrons allow the conduction both of electricity (typical resistivities of 10^{-8} ohm-m) and heat.

SECONDARY BONDS These weak bonds appear because the centre of negative charge due to all the electrons will not, at any given moment, necessarily coincide with the centre of positive charge on the nucleus, the atom will possess a small *dipole moment* at that instant, and a dipole field will surround it. This dipole will fluctuate with time, both in magnitude and direction, depending upon the orbital motion of the electrons. Two neighbouring atoms can therefore interact either through their separate dipole fields or because one atomic dipole can at any instant induce a dipole in its neighbour.

These forces are *very short range and weak* compared with the primary bonds, but they account for the forces which cause rare gases to solidify at low temperatures, and are also of particular interest because they hold together the long covalently bonded molecules to form solid polymeric materials such as polyethylene. They also provide the bonds between the sheets of carbon atoms in the graphite structure, giving it both conductivity and the ability to deform as the sheets slide over each other.

Structure

The metallic bond with its interaction between positive ions and the negatively charged free electron gas forms a bond which has equal attractions in all directions and therefore produces relatively simple atomic structures. The most commonly occurring crystal structures found in metals are shown in Figs. 4 and 5, they are: **face-centred cubic** (f.c.c.), which is adopted by elements such as copper, nickel, aluminium, and important engineering materials such as stainless steels; **close-packed hexagonal** (c.p.h.), which is found in elements such as titanium and magnesium; and **body-centred cubic** (b.c.c.), which is characteristic of mild steel as well as chromium and tungsten.

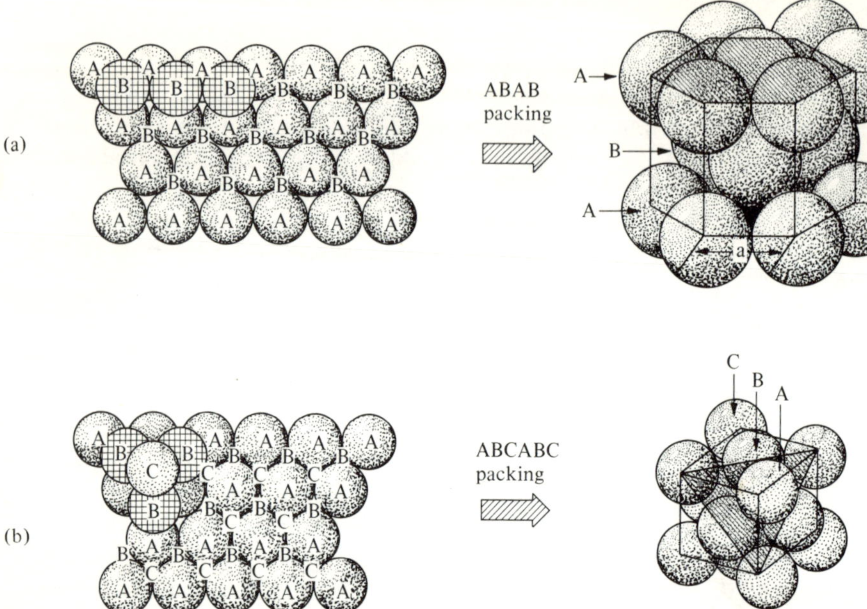

Fig. 4. (a) The development of the close-packed hexagonal structure from the close-packed layer by an ABABAB packing sequence. (b) The face-centred cubic structure developed by an ABCABCABC packing sequence. The close-packed planes are shaded

The face-centred cubic and the close-packed hexagonal represent the *densest* packing obtainable, in which each atom has 12 equally spaced neighbours (i.e. the coordination number is 12). These two structures can be derived from the close-packed layers as shown in Fig. 4. If the first layer is A, then the next layer is B, followed by a third layer which is A again, we shall have generated the stacking sequence ABABAB which is the hexagonal structure. If on the other hand we allow the third layer to become a C layer, as in Fig. 4b, we shall obtain a sequence ABCABCABC – which is the face-centred cubic structure.

Experiment 1. Using table-tennis balls (or polyzote spheres) and suitable glue such as balsa cement, construct a raft of close-packed atoms as shown in Fig. 4. If you make several more of the correct size, preferably coloured with water-soluble ink pens, you will be able to arrange the layers to give either the cubic or hexagonal crystals. Count the number of nearest neighbours (i.e. those in contact with a chosen atom) and satisfy yourself that both structures are the same. Prove in the case of the hexagonal structure that the ratio of the height (c) to the side length (a) is 1·6330.

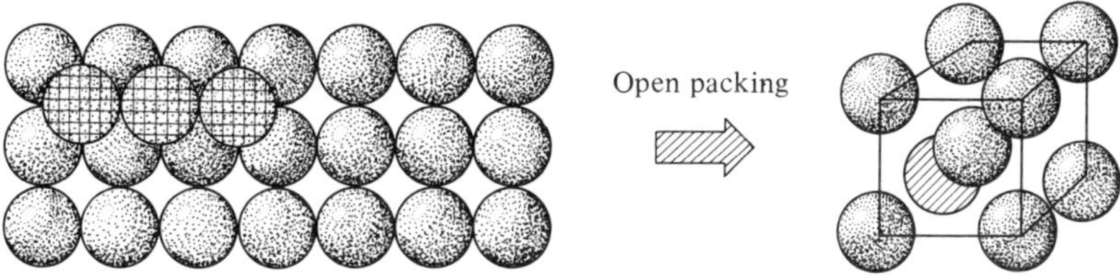

Fig. 5. The open packing of ions leads to the body-centred cubic structure which possesses no close-packed layers

The body-centred cubic structure is derived from the open packing sequence shown in Fig. 5 and produces a unit cell with a coordination number of 8: notice that there are *no* close-packed layers in this structure which leads to a complex deformation under stress, whereas the simpler deformation of the close-packed structures (f.c.c. and c.p.h.) is due to the presence of the unique close-packed planes in their unit cells. These are the base plane in the hexagonal system and the diagonal plane in the cubic system, as shown by the shading in Fig. 4.

It is interesting to compare the packing density produced in each of these packing sequences. The atomic packing factor is defined as the *volume of atoms/ volume of the unit cell* and gives a value of 0·74 for each of the close-packed structures. We might expect the body-centred cubic structure to have a much lower value, but in fact it has a packing factor of 0·68 which is not markedly different from the other two.

Ionically bonded materials, which include most ceramics, crystallize in a wide variety of systems, varying from relatively simple to quite complex. Unlike the metals where simple stacking sequence was the only control on the structure, two important factors have to be considered: (a) the respective ionic valencies establish the ratio of the two atoms that make up the crystal, e.g. Na^+ and Cl^- are present in equal quantities, Ce^{++} and Cl^- are present in the ratio 1:2; and (b) the ratio of the radii of the two ions in a two-component system determines their packing and hence their coordination number. The ions behave as semi-rigid spheres and the stable crystal structure will be the one in which the smaller sphere fits with the least distortion into the voids formed in the packing of the larger ones. Thus a silicon atom with a radius of 0·4 Å* will just fit into the void formed by placing four chlorine atoms of radius 1·8 Å in contact with each other as shown in Fig. 6a. Also illustrated (Fig. 6b) is the case of six fluorine atoms which form an interstitial void of practically the same size as that in the chlorine structure; this leads to the molecular formula SiF_6 as compared with $SiCl_4$ in the previous case.

* The Å unit (which equals 10 nm) is universally accepted by crystallographers.

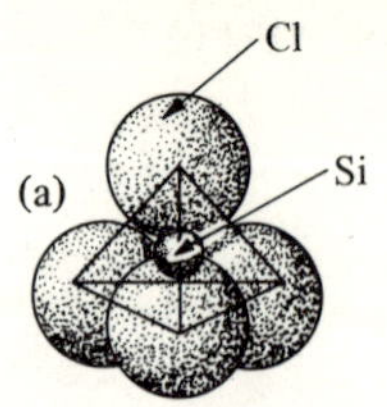

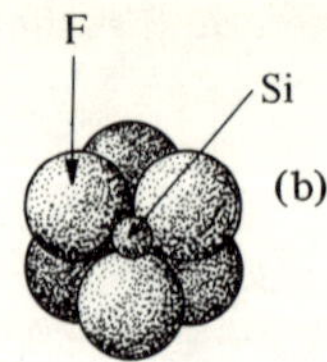

Fig. 6. (a) Four-fold coordination in $SiCl_4$. (b) Six-fold coordination in SiF_6

From this discussion it can be seen that the coordination number may be predicted from the ratio of the radii of the two ions, r_c/R_a such that the anions just touch each other as well as the central cation. For example, in the structure $SiCl_4$ the coordination number is 4 and the ratio is approximately 0·225. The ratios r_c/R_a calculated on the basis of first nearest neighbour are given in Table 1; notice that if the ratio falls between two given values, the lower coordination number will be preferred. It is also interesting to note that the coordination number of 12 observed in the close-packed metal structures may be derived from the same approach.

Table 1 Relationship of atomic radii and coordination number

Coordination No.	Minimum Ratio of Atomic Radii
3-fold	0·155
4-fold	0·225
6-fold	0·414
8-fold	0·732
12-fold	1·000

Experiment 2 Using the data given in Fig. 7 construct a model to show that for a three-fold coordination the ratio of the two radii must be greater than 0·155. Use polyzote spheres for the anions and obtain suitable ball-bearings for the cations.

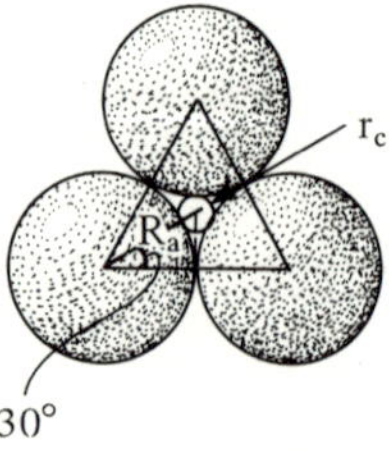

Fig. 7. Minimum ratio of r_c/R_a = 0·155 for a three-fold coordination

Because there are these very definite packing relationships within ceramic materials such as magnesium oxide and aluminium oxide, it is easy to understand why they are *very difficult to deform at normal temperatures*; not only has one to break down the specific anion-cation structure derived from the coordination number, but the deformation must not lead to any change in the electrical neutrality of the crystal by arranging like ions together in a situation where there is a local imbalance in the charge. It is not surprising therefore that most ionically bonded materials will only exhibit reasonable ductility when their temperature is high, thus allowing thermal vibrations in the lattice to assist the deformation process (we shall return to this subject in Chapter 5).

Although ionic materials should possess complex crystal structures, many, such as MgO, CaO, and BaO, crystallize in a face-centred cubic structure as shown in Fig. 8. Notice that this appears to be different from that shown in Fig. 4b. In fact it is the same if it is realized that there are *two* interpenetrating face-centred cubic cells, one based on the *black* atoms and one on the *white* atoms.

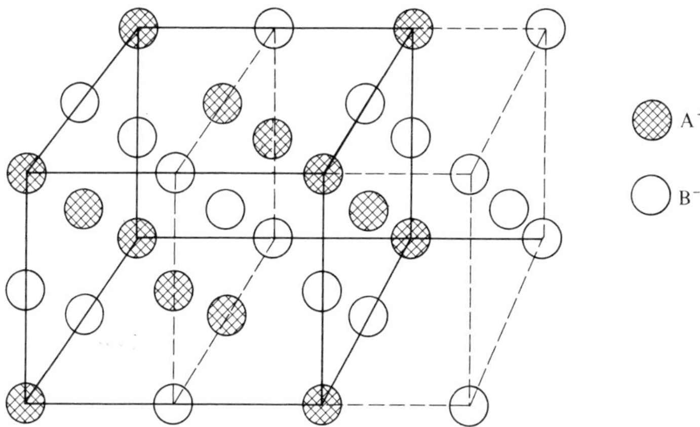

Fig. 8. Unit cell of a A^+B^- ionic crystal (bold lines) showing the interpenetrating f.c.c. (dotted lines) cell

Covalently bonded atoms are found in a number of materials. One example already mentioned is diamond, in which each atom has a coordination number of 4 and therefore the neighbouring atoms are arranged at the four corner points of a tetrahedron. This structure represents the strongest way of arranging these covalent bonds and leads to the observed properties of this form of carbon. The three-dimensional cross-linked polymers such as bakelite and heavily vulcanized rubber are another example of atoms covalently bonded in all three directions. In these polymers however, the three-dimensional structure produces a *supermolecule,* which will not soften on heating as the individual molecules are not free to move; instead, when sufficent heat is supplied to disrupt the primary bonds, the material decomposes.

So far we have confined our discussion to materials which exhibit only one type of atomic bonding; we can now therefore consider those in which *mixed bonds* exist. For example, semiconducting elements such as silicon and germanium possess covalent bonding at room temperature but exhibit semi-metallic behaviour at around 100°C. Although the fascinating subject of semiconduction is outside the scope of this monograph, it is worth noting that the change in behaviour is due to the increased thermal vibrations causing the valency electrons to move away from the covalent bond and become more like the 'free electrons' of the metallic bond. The mixed covalent-van der Waals' bonds that are found in graphite, in which the strong covalent bonded sheets of carbon atoms are held together by the relatively weak van der Waals' bonds, provides another interesting material. The presence of the weak inter-sheet bonds allows them to slide freely over each other thus providing the excellent lubricating properties of graphite. The free nature of the electrons between the sheets gives rise to the low resistivity of the material (10^{-2} ohm-m).

Perhaps the most important materials with mixed bonding are the linear polymers (as shown in Fig. 9), these include thermoplastics such as polythene, polyvinyl

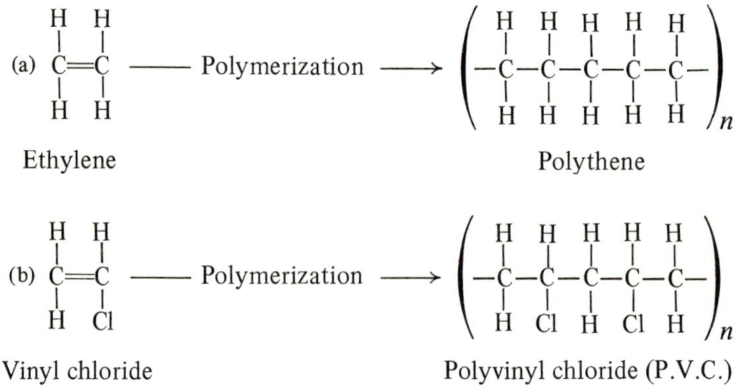

$$\text{(a)}\quad \begin{matrix} H & H \\ | & | \\ C & = C \\ | & | \\ H & H \end{matrix} \;\text{—— Polymerization} \longrightarrow \left(\begin{matrix} & H & H & H & H & H & \\ & | & | & | & | & | & \\ - & C - & C - & C - & C - & C & - \\ & | & | & | & | & | & \\ & H & H & H & H & H & \end{matrix} \right)_n$$

Ethylene — Polythene

$$\text{(b)}\quad \begin{matrix} H & H \\ | & | \\ C & = C \\ | & | \\ H & Cl \end{matrix} \;\text{—— Polymerization} \longrightarrow \left(\begin{matrix} & H & H & H & H & H & \\ & | & | & | & | & | & \\ - & C - & C - & C - & C - & C & - \\ & | & | & | & | & | & \\ & H & Cl & H & Cl & H & \end{matrix} \right)_n$$

Vinyl chloride — Polyvinyl chloride (P.V.C.)

Fig. 9. Polymerization of (a) ethylene and (b) vinyl chloride

chloride, and soft rubber. In these materials the covalently bonded molecules (or monomers) are bound together by van der Waals' forces. The presence of two types of attractive force means that the linear polymers can be developed to possess a much wider range of properties than the cross-linked three-dimensional types, and include materials some of which are soft and rubbery and others which are hard and brittle.

There are three important factors which govern the rigidity of these polymers.

(1) *The degree of overlapping* of the molecules which is proportional to the formula weight and is normally referred to as the degree of polymerization (N); there being a characteristic relationship between the strength and N for all linear polymers.
(2) *The orientation of the polymer chains,* which in some senses is equivalent to the packing of ionic sheets in metallic crystals. In the molten state, polymer molecules are randomly orientated. If the solution is quickly frozen this random

orientation is maintained and an **amorphous** polymer results. If, however, the molten polymer is allowed to cool very slowly, small regions of the structure can exhibit **crystallinity**, that is there is a definite arrangement of the polymer chains (as sketched in Fig. 10c). If the crystalline polymer is mechanically cold-worked the crystalline regions can be drawn out and an orientated crystalline polymer results (Fig. 10d).

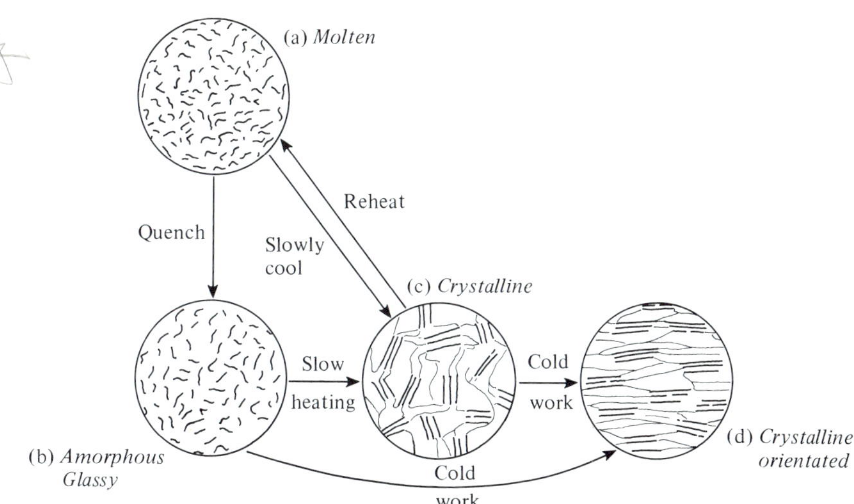

Fig. 10. The structure of a linear polymer in different conditions

(3) *The degree of cross-linking* and cross-branching between the chains will also affect the rigidity, by making it difficult for the polymer chains to move and rotate to accommodate the applied stress. This explains why polythene produced at high pressures (which causes chain-branching during the polymerization reaction) is softer than the highly crystalline material produced at low pressures, which because of the lack of branched chains is free to orientate itself under stress and therefore to produce a harder material. We shall have to bear these important structural aspects in mind when we come to consider the mechanical properties of polymers and rubbers in Chapter 4.

Finally in this very brief survey of structure and bonding let us consider the material, glass. Experimental data suggest that it is bonded by covalent forces within the silica molecules but has ionic bonds to produce electrical neutrality, and is similar in many ways to a *supercooled liquid* (the glassy or amorphous polymer is probably similar). In order to understand the structure of glass we can consider the model (Fig. 11a) for silicon with a four-fold coordination with oxygen. The tetrahedra could be rigidly joined together in a true crystalline structure with the bond angle fixed, but if we allow the angle between the tetrahedra to vary, then a random structure will result (Fig. 11b). Although the silica tetrahedron is the basis for the common glasses, additional large cations are also present so that the general formula for glass is $A_m B_n O$ where A is a large cation with a small charge such as Na^+, B is a large tri- or tetravalent ion such as B^{3+} or Si^{4+}, and O is oxygen.

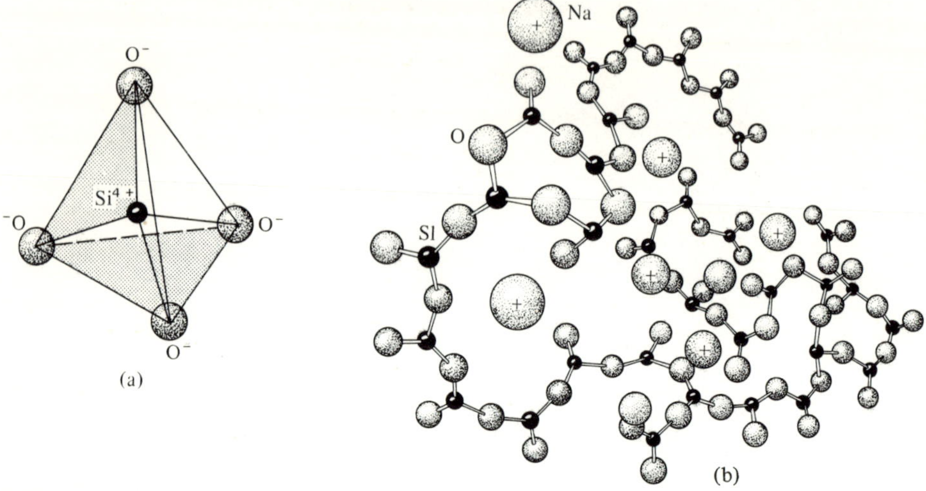

Fig. 11. The structure of (a) a simple silica tetrahedron and (b) sodium glass

The B and O ions are covalently bonded to form a continuous chain. The A ions which serve to produce an electrical neutrality in the structure can migrate through the glass without affecting the chain network, and it is this freedom that allows the glass to become viscous without breaking down at elevated temperatures.

3 Metallic materials

Pure Metals and imperfections

We saw in Chapter 2 that the metallic bond led to the unique situation of a set of positive ions embedded in a free electron gas and we might therefore expect a metal under stress to deform in a quite random manner under stress as the ions can move equally well in any direction. This is not the case and an examination of a metallic crystal will reveal that deformation (or slip) occurs along *specific crystal directions* and on *specific crystal planes.* If we consider the analogy of friction, the plane which is the smoothest atomically should be the preferred slip plane, i.e. deformation should occur more easily on that plane than on any other. This is certainly true for the crystal structures that have close-packed lattices, namely the close-packed hexagonal and the face-centred cubic. In Fig. 4 the close-packed layers are shaded and deformation occurs by the movement of one layer over another which results in the production of slip bands on the surface of the crystal (both a schematic model and a real photomicrograph are shown in Figs. 12a and 12b).

Experiment 3. Using two rafts of close-packed polyzote spheres, fix one down to a bench and slide the top layer very carefully over the other and notice that the movement is not a simple linear translation but a zig-zag movement, due to the fact that this is the path of least resistance in the lattice. From your model calculate the actual distance moved to produce a given total displacement of the layers.

The force required to move one layer over another should be a measure of the theoretical strength of a metal. Consider the model shown in Fig. 12c and notice the periodic variation in the stress (τ) as a function of the distance (x) that the plane moves.

If we can assume that τ varies according to sine function (although other assumptions are possible and indeed desirable for improved accuracy of the shear stress estimate), the value of τ is equal to $C/\sin(2\pi x/a)$, where C is constant and x is the shear displacement or the distance over which the upper atom moves with respect to the lower one. The value of the constant C is found to be $\mu a/2\pi h$ where μ is the shear modulus (i.e. shear stress/shear strain) and thus our stress equation becomes:

$$\tau = \frac{\mu a}{2\pi h} \sin\left(\frac{2\pi x}{a}\right) \tag{2.1}$$

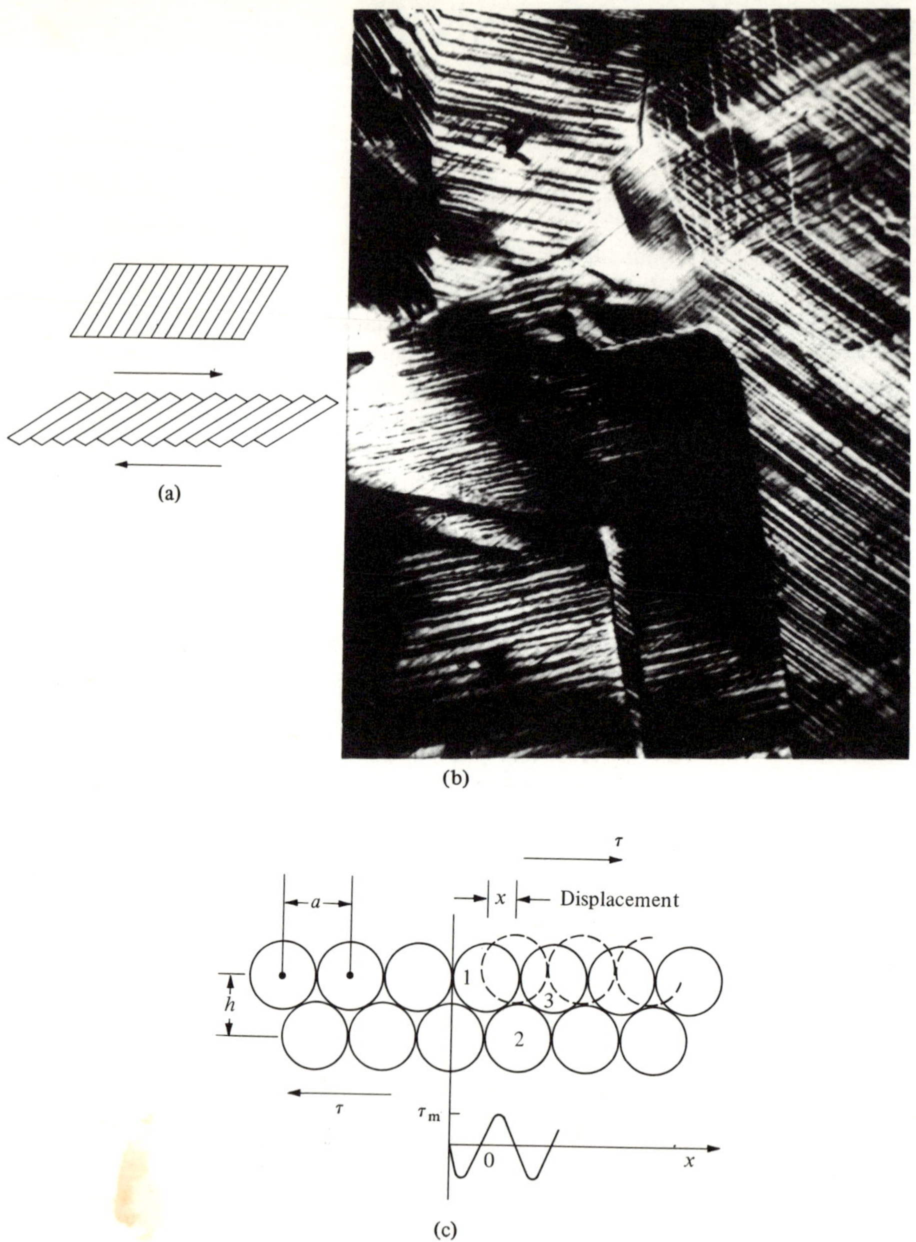

Fig. 12. The mechanism of slip in metal crystals. (a) Early interpretation of the slip patterns observed in (b). In (b) notice how the slip planes vary from one crystal to the next. (c) Theoretical model of slip

The value of τ will be a maximum at $\sin(2\pi x/a) = 1$, i.e. at $x = a/4$, hence the lattice becomes unstable and slip will occur when:

$$\tau_m = \frac{\mu a}{2\pi h} \qquad (2.2)$$

Since both a and h are of the same value (typically 3 to 5 Å) the theoretical deformation stress should be $\sim \mu/5$. Typical values of μ for mild steel are 68 GN m^{-2} and therefore deformation should occur at approximately 13 GN m^{-2}. In practice the observed value is rarely higher than $\mu/1000$ (0·05 GN m^{-2}). This *extraordinary disparity* must force us to revise our thinking about the mechanism of slip.

Experiment 4. Cut a small cylinder or cube from a piece of pure copper and then carefully polish a face on the block using suitable emery cloths and then metal polish. The grain structure can be observed if the polished face is etched for 15 to 30 seconds in 5% iron (III) chloride/2% hydrochloric acid in alcohol. Using a similarly polished block, compress it in a vice some 5% or in a tensile machine so that a slight rumpling occurs on the polished face. If this face is now viewed at x10 to x30 magnification, the various slip bands may be observed (similar to Fig. 12b). Notice how the slip lines change orientation from one crystal to another.

To account for the relative ease with which slip takes place in crystals it is necessary to assume that they contain certain structural defects, the most important of which is **the dislocation.**

If we are to understand the mechanical properties of metals and their alloys, we must understand how this defect moves, can be stopped and controlled in the crystal lattice. The dislocation is very mobile when left to itself and the whole art of making metals stronger is to block the dislocation so that it cannot move so freely. First let us look at a model of a crystal containing a dislocation. One of the simplest and *most important forms* is shown in Fig. 13a.

This is an *edge dislocation* and it consists of an extra half plane of atoms stretching into the paper which has resulted from a deviation in the packing sequence which built up the crystal structure. Notice that owing to the elastic nature of the metallic lattice the dislocation is accommodated by the distortion of the lattice lines (i.e. the lines joining atom centres) around the defect. We can use a simple illustration to show how such a dislocation enables one plane of atoms to move bodily over another at stresses much smaller than those needed to break all the bonds between the planes simultaneously. Consider the practical problem of moving a carpet some distance across a floor; the carpet may be dragged bodily across the floor, but this requires a large force. A much easier way is to form a linear bulge across the carpet by rucking up one end of it and pushing the ruck from one end to the other. As the ruck sweeps along the carpet, it easily shifts the

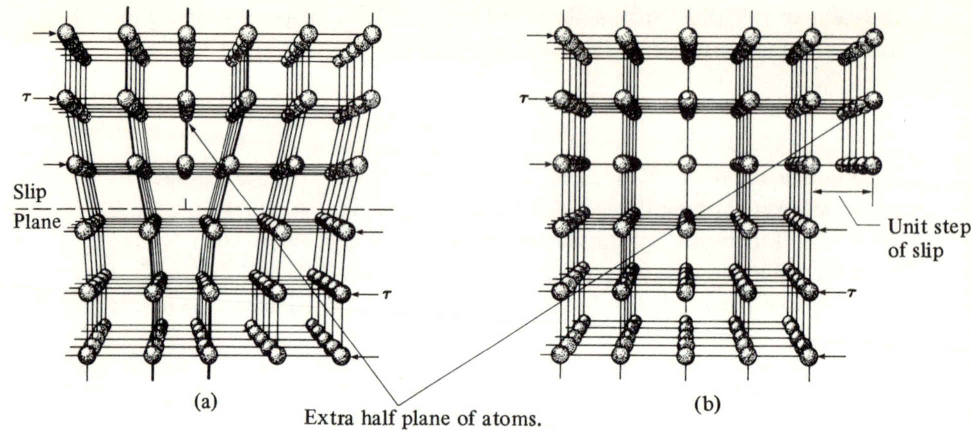

Fig. 13. A dislocation in a simple cubic structure; notice the extra half plane in (a) and how it will produce a slip step when moved by a suitable shear stress

whole carpet a distance equivalent to the original ruck. The dislocation may be observed experimentally using an electron microscope and much of our understanding of metallic behaviour under stress has stemmed from these detailed observations; Fig. 14 shows a group of dislocations forming a ladder network in a thin foil of a high-strength magnesium alloy.

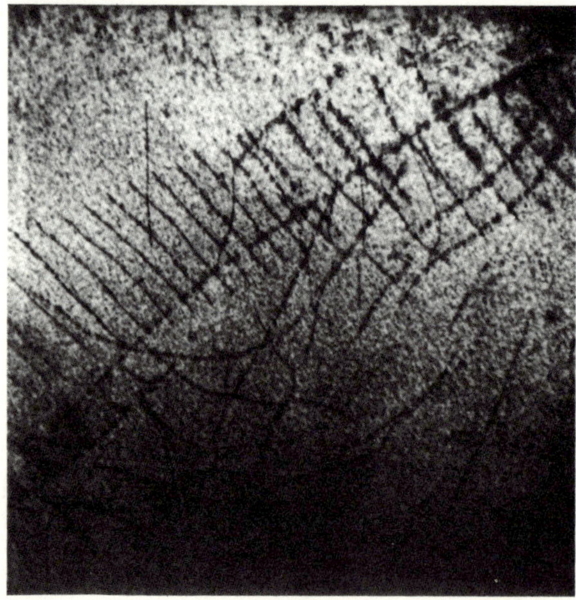

Fig. 14. Ladder network of dislocations in a thin magnesium foil at ×20 000.

Experiment 5. There are two methods of illustrating the formation and behaviour of dislocations in two dimensions. (a) A projection slide (preferably 70 mm x 70 mm) is made so that it can be filled with very small ball bearings (about 1 to 1·5 mm diameter). If this is mounted in a projector various effects can be produced including packing sequences, crystal boundaries, vacancies, and dislocations. Dynamic effects can also be achieved if the slide is lightly tapped on one edge. The effects of alloying can also be illustrated by adding one or two oversized ball bearings so that strain distortions are introduced into the packing. (b) A more elegant though more difficult technique is to form a raft of soap bubbles on the surface of a soap solution contained in a back-lit translucent box. Similar effects as in (a) can be observed and the dynamic behaviour studied if the raft is carefully compressed or stretched. (It is strongly suggested that the film 'The Bubble Model of a Metal Structure'* is seen before attempting to make a bubble raft.)

Stress-strain curves

The commonest method of expressing the mechanical properties of a material is by a *tensile stress-strain curve,* in which the stress (or load) on the sample under test is plotted as a function of the strain (or extension). For further details of this test the reader is referred to the appendix which covers not only the tensile test but compression, bending, fatigue, creep and hardness testing. Fig. 15 illustrates the typical results one might expect to obtain from a range of different metallic materials; notice the distinctive difference in the shape of the curves obtained for body-centred cubic mild steel and face-centred cubic copper.

Consider the curve obtained for pure copper, after the initial elastic portion (i.e. up to *P*) in which load was proportional to extension (*the constant of proportionality E is termed Young's modulus*). The material deforms plastically with a gradual increase in the load, and it becomes increasingly difficult to deform the copper further. This is known as *work hardening* and it can be easily demonstrated that if the load is reduced to zero at point *A*, on retesting the sample the material will remain elastic until the load has reached the original value from which it has been reduced (point *B*). The shift along the extension axis is due to *elastic hysteresis.* After the load has reached the maximum at T_c, a neck is formed in the sample which weakens it at that point and the load begins to fall before final fracture at point *F.* The maximum load prior to fracture is called the *ultimate tensile load* (or stress if divided by the original cross-sectional area of the specimen). The two fractured halves of a copper sample are shown in Fig. 16 in which it is easy to see the necked-down area and the typical cup and cone fracture surface of a ductile fracture.

* Obtainable from: MacQueen Film Organisation, West Street, Bromley, Kent.

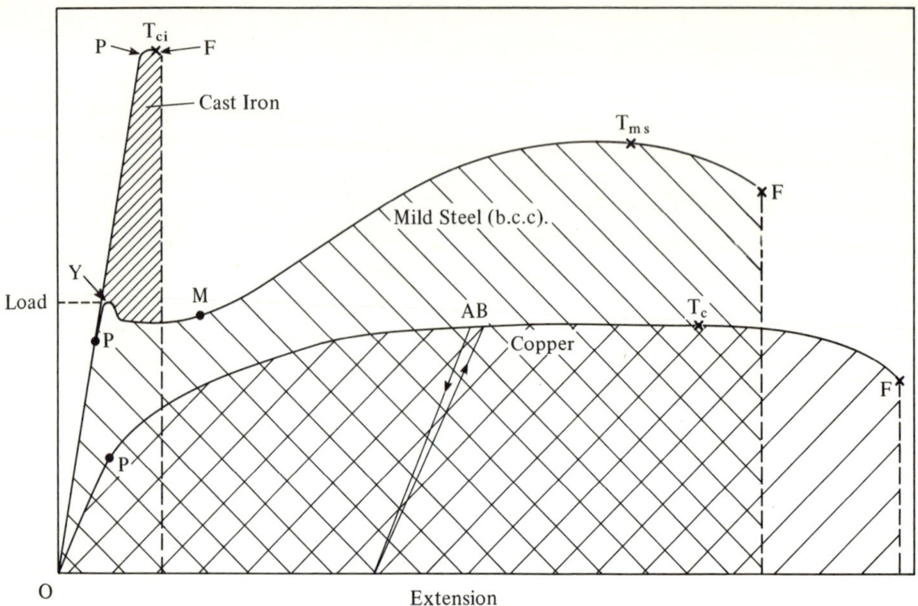

Fig. 15. The load-extension curves for typical face-centred cubic metals. The areas under the curves are measures of the modulus of toughness

In the case of mild steel, the material behaves as an elastic solid up to the point *P*, but beyond *P*, although the material continues to behave in an elastic manner, it no longer obeys Hooke's law. A *yield point* is reached at *Y* which represents the stress needed to produce numerous mobile dislocations in the material and the number of dislocations increases from about 10^{10} m^{-2} to some 10^{16} m^{-2} just prior to fracture. The release of these mobile dislocations in a steel accounts for the ductile region

Fig. 16. The typical cup and cone fracture produced by a ductile metal

$Y \rightarrow M$ in which the load *hardly increases.* The onset of work hardening at M causes the load to rise as it becomes increasingly difficult to deform the steel further. Once again the load falls from the maximum T_{ms} due to necking of the sample prior to fracture at F.

Cast iron is an interesting material as it normally behaves as an elastic solid almost to the point of fracture, therefore the stress-strain curve is very simple, with the points P, T_{ci}, and F hardly distinguishable. If the toughness of the metal is defined as the area under the load-extension curve, we may clearly see the distinction between toughness and strength, cast iron is stronger than either mild steel or copper (i.e. $T_{ci} > T_{ms} > T_c$) but it has a much lower toughness; this distinction is so important that the majority of the effort in advanced materials technology in rockets and aerospace environments is devoted to the production of very strong and very tough materials.

Experiment 6. If you have a suitable testing machine (one that is able to measure both load and extension), use thin sheet specimens (approximately 1 to 2 mm thick) of annealed copper (i.e. heated to 300°C for 1 hour), mild steel and aluminium to plot the load-extension curves for each of the materials at three different temperatures: 100°C, room temperature, and – 77°C (alcohol saturated with dry ice). Compare the fracture surfaces of the specimens and notice how copper and aluminium become more ductile as the temperature is lowered. If you have access to liquid nitrogen you will be able to plot the results down to – 120°C (approximate melting point of alcohol). The design of a suitable temperature enclosure is briefly discussed in the appendix.

We might at this point ask how we can view the *experimental stress-strain curves in terms of the dislocations* that exist in the material. As a result of modern research in just this area, we are able to explain the experimental curves quite well. In the Hookean region, the dislocations begin to move from their initial sites in the lattice, but return elastically to that site when the stress is removed. At loads between P and Y some of the dislocations move a small distance from the original site, so that on the removal of the load there is a small but finite plastic strain in the material. Once the point Y is reached for mild steel, or the point P for copper, there is a large movement of the total dislocation population which causes the large plastic strains measured; if one dislocation produces a displacement of 5 Å then at least $10^7 \rightarrow 10^8$ are required to produce 1 mm of slip. One might expect the dislocations to continue to move at a constant load until the material fractures, but although such behaviour has been observed in single crystals of hexagonal metals deforming on their close-packed planes, the usual situation is for a progressive increase in the load to maintain the plastic strain (i.e. between P or M and T). The explanation of work hardening is based on the model of mobile dislocations becoming tangled up with crystal boundaries and, more importantly, *with themselves.* As these tangles (which are rather like crystalline traffic jams) become increasingly severe, the load required to keep dislocations moving rises to a maximum (at the point T) before fracture occurs.

The hardening of polycrystalline metals

ALLOY HARDENING Pure metals are relatively weak and therefore their use for practical purposes is limited unless it involves the use of another property such as the low resistivity in copper, the high corrosion resistance of gold or the high melting point of tungsten; but where *strong materials* are required we must make use of *alloys.*

If the plastic behaviour of these metals and alloys is controlled by the movement of dislocations, then the production of stronger and better materials will depend upon the control of their movement together with attempts to block the movement completely.

The oldest method of achieving this is to produce an alloy by introducing foreign atoms into the lattice whose size, valency, or elastic modulus is different.

Fig. 17. Solute atoms can enter into a parent metal to form an alloy by substitution (B) or interstitial formation (C)

Alloys may be divided into two important classes as shown in Fig. 17: (i) where the new atoms fit into the existing lattice sites (strictly speaking the element enters the alloy in the ionic state), as in stainless steel (Fe–Ni–Cr), brass (Cu–Zn), and bronze (Cu–Sn); or (ii) where the new atoms, usually because of their small size, fit into interstitial sites in the lattice, the best-known example being mild steel (Fe–C). Because the *alloy atoms produce a distortion in the original lattice* the mobile dislocations will require extra energy to force their way through and hence the material will be much stronger and harder. For example, the ultimate tensile strength of pure iron is 80 MN m^{-2}, but steel containing 0·2% C has an ultimate tensile strength of 460 MN m^{-2}. Similarly, the tensile strength of pure copper is increased from 50 MN m^{-2} to 100 MN m^{-2} by the addition of 30% Zn to form brass Fig. 19. Let us think a little more about our two examples; the addition of zinc to copper is a *simple substitution* in the copper lattice, and one can continue adding zinc until the solubility limit is reached at about 37 wt % Zn (like a saturated chemical solution). If more zinc is added, the copper lattice is so distorted that it cannot accommodate any more zinc atoms in solution and a new zinc-rich phase (called β) is formed, so that the alloy now contains two phases $\alpha + \beta$ as seen in Fig. 18b.

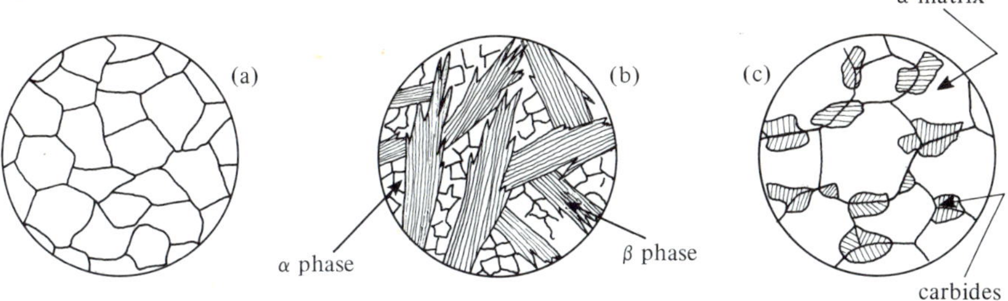

Fig. 18. Microstructures of (a) α solid solution brass, (b) duplex $\alpha + \beta$ brass, and (c) mild steel

This behaviour is typical of substitutional solid solutions; the behaviour of interstitial solid solutions is similar, but owing to the *high strain* introduced into the lattice by the interstitial atom the solubility is normally limited to a *few per-cent* (0·003 wt % in the case of C in iron at room temperature). The excess carbon is present as a second phase *cementite* (Fe_3C), and it is this very hard and brittle phase that strengthens steel rather than the interstitial solid solution matrix.

Experiment 7 Using hardness as a measure of the tensile strength (hardness is often proportional to the ultimate tensile strength) examine the effect of alloy additions on the mechanical properties of various pure metals, such as Cu, Cu–Zn, Cu–Ni, Cu–Sn, Fe, Fe–C, Pb, Pb–Sn. Estimate the efficiency of the strengthening addition in each case as the increase in hardness/unit % of alloy addition. (Hint: In obtaining the material it must be specified as being required in the annealed or fully softened form, other forms of the material will affect your comparative results.)

Experiment 8. Using a single crystal (grown if possible in your laboratory) of either zinc or tin, measure the hardness and the ultimate tensile strength. Compare these results with those obtained on a rod of similar dimensions which is polycrystalline (these are easily produced by pouring the molten metal into a copper or iron mould to induce rapid cooling). You will find that the presence of the grain boundaries (these are easily seen if the surface of the rod is swabbed with 5M HCl) causes the material to be much stronger owing to the dislocation barriers produced.

We can now consider another important type of alloy hardening mechanism, known as *precipitation or age hardening.* Consider the phase diagram shown in Fig. 20 (phase diagrams indicate the state of the material as a function of composition, temperature, and pressure; the pressure is usually taken as 1 atmosphere). If the alloy containing $x\%$ B is cooled very slowly from the molten state, it will have a structure similar to the $\alpha + \beta$ brass shown in Fig. 18b, the strengthening being produced by the hard second phase θ.

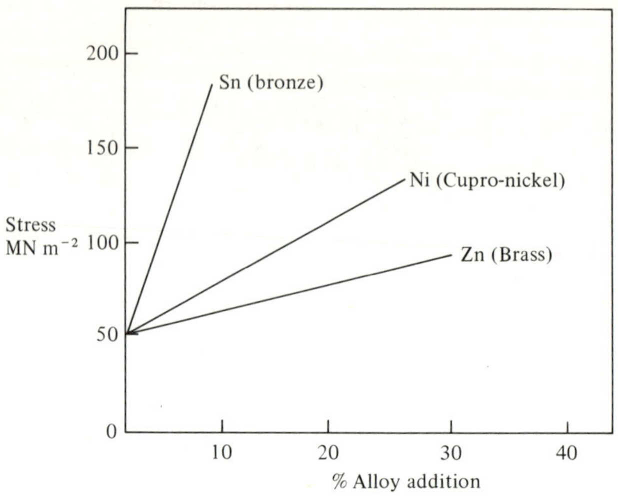

Fig. 19. Effect of solute alloy elements on the elastic limit of various polycrystalline copper alloys

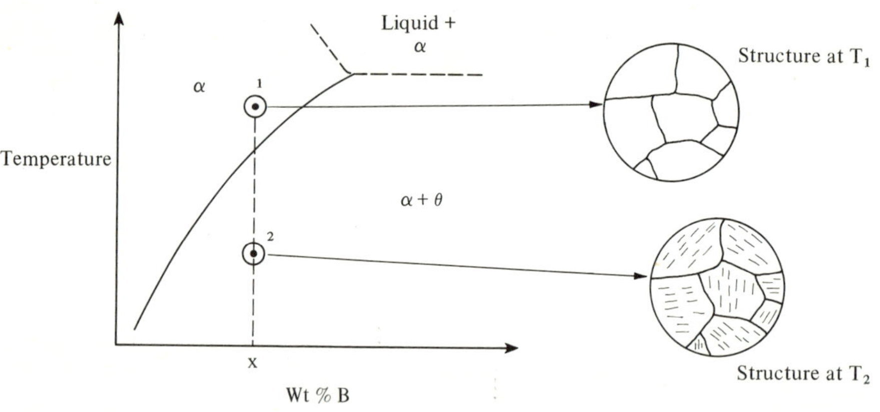

Fig. 20. Phase diagram of a binary alloy which exhibits precipitation hardening. Point (1) represents the solution treatment temperature and point (2) the ageing temperature.

It is possible, however, to produce a much stronger material by controlling the way the θ phase is precipitated in the matrix of α. If the alloy is heated up to temperature T_1 the θ phase will dissolve and we shall have a complete substitutional solid solution. The material is rapidly cooled to room temperature, which retains the B atoms in solution (i.e. a supersaturated solution). By heating the material at a temperature T_2 (typically 130 to 200°C in Al–Cu alloys) the second phase is re-precipitated in a *very fine metastable distribution* of θ' and θ'')* before the equilibrium θ appears. Because these metastable precipitates are so fine (they cannot be observed with an optical microscope), they easily entangle the dislocations and so produce a very hard material. Obviously the time of heating at the precipitation temperature is important and the variation of hardness with time can be seen clearly in Fig. 21.

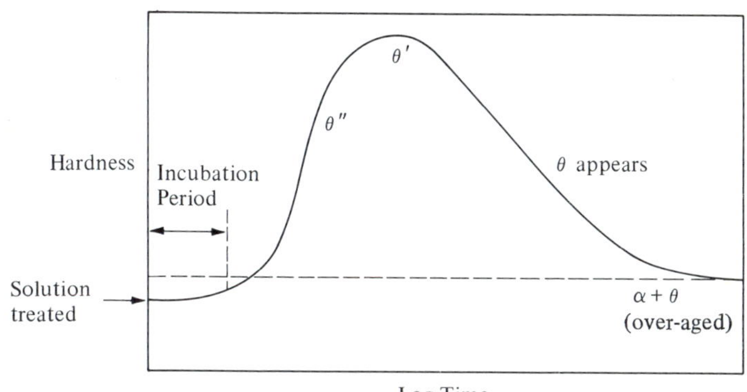

Fig. 21. The variation of hardness with time for an age-hardened alloy held at the same temperature

Experiment 9. Using an aluminium 2 → 4% Cu alloy (duralumin), the results shown in Fig. 21 can be obtained if the alloy is solution-treated at 525°C for 1 hour and quenched into water, followed by ageing at 180 to 200°C for times up to 3 hours. The mechanical properties are best observed using hardness or a simple elastic limit measurement. If it is possible compare the microstructures of alloys which have been solution-treated, aged to peak hardness and over-aged.

Perhaps the most striking illustration of the ability to develop strength by this method is found in the copper-2% beryllium alloys in which the ultimate tensile strength may be increased from 15 MN m^{-2} to 2000 MN m^{-2}. This aged material is strong enough to be made into drills and chisels with which to cut ordinary steels.

WORK HARDENING Most people will have experienced the effect of work hardening as they have bent a piece of wire between their hands, in that it becomes *increasingly difficult* to bend the wire before final fracture occurs; they are in fact stressing the material in the portion *M* (or *P* for copper) → *T* on the stress-strain curve. The explanation of this effect once again involves dislocations. As the mobile

* θ' and θ'' are pre-precipitates which are crystallographically related to the matrix.

dislocations begin to move at the elastic limit they soon become entangled with one another and with the grain boundaries, this in turn raises the internal energy of the lattice and causes further dislocation movement to become difficult. This is therefore a very simple but efficient method of raising the elastic limit of a metal, as illustrated in Fig. 22 (steel reduced to 2% of its original area in the form of piano wire has an elastic limit of 4 GN m^{-2} which approaches the theoretical value), but it has little effect on the ultimate tensile strength, since as these two values approach each other, the material unfortunately becomes more brittle and less attractive to use.

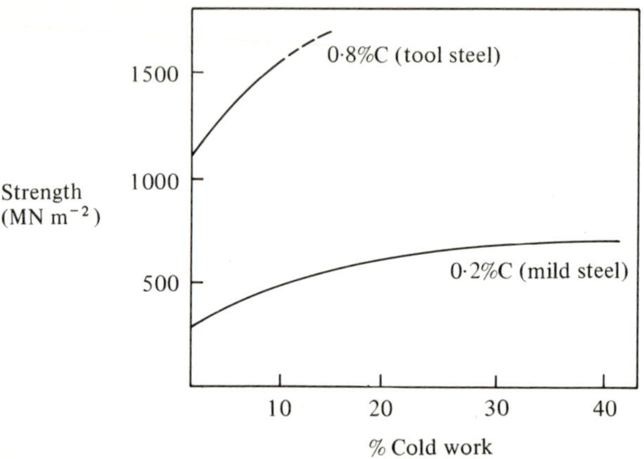

Fig. 22. Effect of the amount of previous cold-work on the subsequent ultimate tensile strength of mild steel and a high carbon steel

However, in mild steel for example, which is to be used as reinforcing bars in concrete, moderate amounts of cold-work can improve the overall strength. Cold-worked metals can be returned to their original state if a suitable heat-treatment is given (annealing at 0·25 of the melting point in °C), which allows the dislocation tangles to sort themselves out and eventually to disappear.

CRYSTALLOGRAPHIC CHANGES Finally in our discussion of strengthening mechanisms let us consider the method that may be applied to metals which exhibit an allotropic change. The best-known example of this is iron, which transforms from an a f.c.c. (γ) to a b.c.c. (α) structure at 910°C in pure iron and at 723°C in a steel containing 0·8%C. If mild steel is cooled slowly the microstructure will appear as in Fig. 18c with a very fine distribution of carbide particles around the grain boundaries of the α (b.c.c.) phase. If this steel is quenched from 850°C to room temperature, there is insufficient time for the carbides to be precipitated from the high-temperature solid solution and we shall obtain a *supersaturated solid solution* (called martensite) as in the case of the aluminium-copper alloys, but with two important distinctions: (i) the iron will undergo the allotropic modification from $\gamma \rightarrow \alpha$; and (ii) the carbon atoms in the iron are in an interstitial solution and will induce very large distortions into the lattice when the iron is quenched. The martensite will therefore be very brittle and hard (the actual hardness depending on the percentage of carbon in the steel). As with precipitation-hardened materials, the mechanical properties of the steel can be improved if the martensite is aged (called *tempering*) to produce a very fine dispersion of the iron carbide particles in the α matrix. This ability to alter the properties of iron over a wide range, together with the availability of its ores, is the reason why iron and its alloys are the most widely used materials for most engineering purposes.

Experiment 10. A steel pin can be used to demonstrate the various properties of this material. If the pin is heated in a bunsen flame and allowed to cool in air, it becomes very soft and is easily bent, but if it is quenched rapidly in water after heating it cannot be bent and breaks in a brittle manner. The ductile state can again be induced by reheating the quenched pin at a temperature which just causes a blue oxide colour to appear on the polished surface (i.e. tempering).

Failure of metals

FRACTURE We saw earlier in this chapter that all metals will eventually fail when the load is sufficiently high and, although the fracture mechanisms in most metallic systems are quite complex, they usually fall into two distinct groups: (i) *transcrystalline,* where the fracture occurs randomly across the crystals; and (ii) *intercrystalline,* where the fracture occurs along the crystal boundaries. Temperature is probably the most important control on the fracture mechanisms. Mild steel at room temperature is ductile and the fracture occurs along the planes of maximum shear stress to produce the typical cup and cone fracture as in Fig. 16; but if it is tested at – 150°C the steel is brittle and, although the fracture is still transcrystalline, it has occurred by cleavage across specific crystal planes (usually the faces of the body-centred cube). Face-centred cubic metals on the other hand fail in a ductile manner at all temperatures and at very high temperatures the specimen will reduce its cross-sectional area to the minimum before it simply pulls down to a point and parts. The transition from brittle to ductile failure is of paramount importance in the structural use of materials, and simple tests have been devised to detect the temperatures at which brittle fracture will occur. In the *Charpy test* a square-section specimen containing a sharp notch is held rigidly in a vice and struck by a swinging pendulum, the energy absorbed being noted as a function of the temperature. When brittle failures occur, the energy (in Joules) falls rapidly, as may be seen from Fig. 23, and the transition temperature is quoted for some particular standardized value of absorbed energy (normally 20 J). The change in the fracture nature is clearly illustrated in Fig. 24 for a mild steel which has a transition temperature of approximately 25°C.

FATIGUE Materials can be subjected to many types of loading, but perhaps the most severe is the *oscillating or fatigue load.* Most people will have heard of aircraft being grounded owing to the premature failure of some part in fatigue, or of a broken

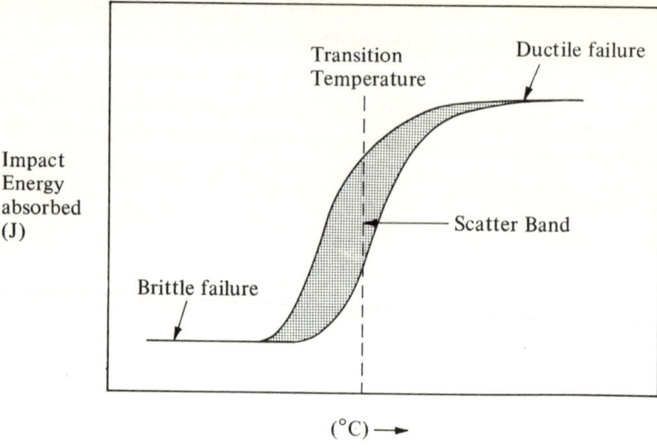

Fig. 23. Impact energy-temperature curve (Charpy test) for a mild steel

Fig. 24. Appearance of the fracture faces of impact specimens broken at –13°C, 12°C, 62°C, 97°C, and 117°C

car back-axle. The fatigue strength of a material is its ability to withstand the oscillating stress for a very long period of time (or a large number of reversals) and is measured by performing a test in which the cycles to failure are measured for specific mean loads; Fig. 25 illustrates typical results for steel and aluminium. The reason why metallic materials fail in fatigue at stresses well below their elastic limit is quite complex, but we may notice two points: (i) the material's ability to work harden is *not realized under oscillating loads*; and (ii) the ease with which *surface cracks* due to abrasion or corrosion may *propagate* through the material.

It is important to notice that steel exhibits a limit of stress (endurance limit) below which the material will not fail in fatigue within a very large number of cycles. Aluminium on the other hand does not appear to reach a limit and the fatigue life gradually increases as the stress drops. Although, from Fig. 25, it would appear that aluminium has a poorer fatigue behaviour, this is not so, as the comparison between the two materials should be made on the basis of fatigue stress (at a given number of cycles)/ultimate tensile strength.

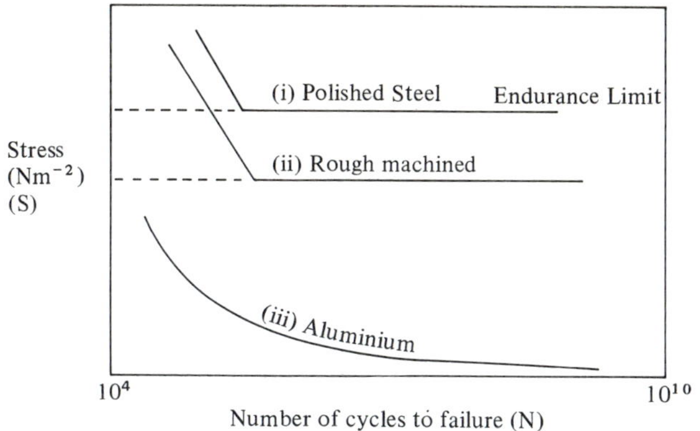

Fig. 25. The S-N curves for: (i) polished steel; (ii) rough machined steel; and (iii) aluminium

Experiment 11. You may construct a fatigue machine using a simple motor which can run at a constant known speed, fit it with a drill chuck which will accept 5 to 8 mm diameter rods. The load may be applied by fitting to the other end of the rod a ball-bearing race to which is attached a weighing pan (see Fig. 40). If the rod is reduced in section in the centre it will ensure that failure occurs outside the gripped areas. Start with a very large load that will cause failure within a few thousand cycles and then gradually reduce the load to obtain the S-N diagram. Various surface finishes can also be tested; notice that specimens containing sharp angles, surface cracks, or poor surfaces have a much lower endurance limit than a polished sample.

4 Polymeric materials

Introduction

A large group of engineering materials of increasing importance are the natural and synthetic high polymers. These materials range from soft rubber to hard and glassy thermosetting resins. As with metallic materials, a knowledge of the internal structure is necessary to understand the extremely complex mechanical behaviour observed. Being organic compounds they are not resistant to high temperatures and easily burn and decompose, but their mechanical properties are *infinitely variable* so that they can be 'tailor made' for specific applications. A selection of these materials is given in Table 2.

Structure and crystallinity

Unlike the metallic materials we considered in the previous chapter, the mechanical properties of polymeric materials are *extremely sensitive to their structure, the temperature of testing and to the way in which they are tested*; therefore in order to understand their complex behaviour we must consider their structure in more detail.

In Chapter 2 we saw that polymers can be produced in two distinctive forms: the *amorphous* (or glassy) and the *crystalline* in which the polymer chains are arranged so that there are areas which are similar to the regular packing found in metallic crystals. Consider the variation of the *specific volume* (volume per unit mass) of the two types of polymers as a function of temperature as illustrated schematically in Fig. 26.

In the case of the amorphous polymer (e.g. polyvinyl chloride) in which the polymer chains are arranged in a purely random manner, there is little or no change in the specific volume at the freezing point of the liquid (T_m) and the polymer retains a structure below this temperature ($T_m > T > T_g$) which is similar to a supercooled liquid such as glass. In this state the polymer chains still have a certain freedom of movement relative to each other; however, below the temperature T_g (known as the **glass transition temperature**) the polymer chains become locked and little relative movement occurs. Strictly speaking T_g is not a *fixed temperature, but a temperature range*; however, specific values are quoted because the relative position of room temperature to this value controls the mechanical behaviour of the polymer at normal temperatures. So polymers which have $T_g >$ room temperature (T_r) will be hard and brittle, whereas those with $T_g \approx T_r$ will exhibit viscoelastic properties, whilst those with $T_g < T_r$ will behave more like rubbers and exhibit rubber elasticity.

Table 2

Material	Polymer Unit	State at room temp.	*UTS* ($MN\,m^{-2}$)	*E* ($MN\,m^{-2}$)	T_g (°K)
Polyethylene	$[\bullet\!-\!CH_2\!-\!CH_2\!-\!\bullet]_n$	Crystalline	7–14	70–280	153
Polyvinyl chloride	$[\bullet\!-\!CH_2\!-\!CH(Cl)\!-\!\bullet]_n$	Amorphous/ slightly crystalline	28–40	2500–3500	353
Polystyrene	$[\bullet\!-\!CH_2\!-\!CH(C_6H_5)\!-\!\bullet]_n$	Amorphous/ crystalline	35–50	3500–4200	373
Nylon 6:6	$[\bullet\!-\!NH\!-\!(CH_2)_6\!-\!NH\!-\!C(=O)\!-\!(CH_2)_4\!-\!C(=O)\!-\!\bullet]_n$	Crystalline	50–70	2000–3000	323
Polymethyl-methacrylate	$[\bullet\!-\!CH_2\!-\!C(CH_3)(O{=}C\!-\!O\!-\!CH_3)\!-\!\bullet]_n$	Amorphous	50–70	2500–4000	380
Polytetrafluor-ethylene	$[\bullet\!-\!CF_2\!-\!CF_2\!-\!\bullet]_n$	Crystalline	14–30	400–650	399
Phenol formaldehyde resin	$[\bullet\!-\!C_6H_2(OH)(CH_2\!-\!\bullet)\!-\!CH_2\!-\!C_6H_2(OH)(CH_2\!-\!\bullet)\!-\!CH_2\!-\!\bullet]_n$	Thermosetting glass	50	7000	–
Polyisoprene (natural rubber)	$[\bullet\!-\!CH_2\!-\!C(CH_3){=}CH\!-\!CH_2\!-\!\bullet]_n$	Elastomer	2–10	7–70	203

For crystalline polymers there are two distinct changes in the specific volume curve: (i) the glass transition at T_g; and (ii) the melting point of the crystallites at T_m. In highly crystalline polymers there is a very *large change in volume* at this temperature (≈ 10%) which is due to the packing together of the polymer chains in constrast with their random distribution in the liquid state.

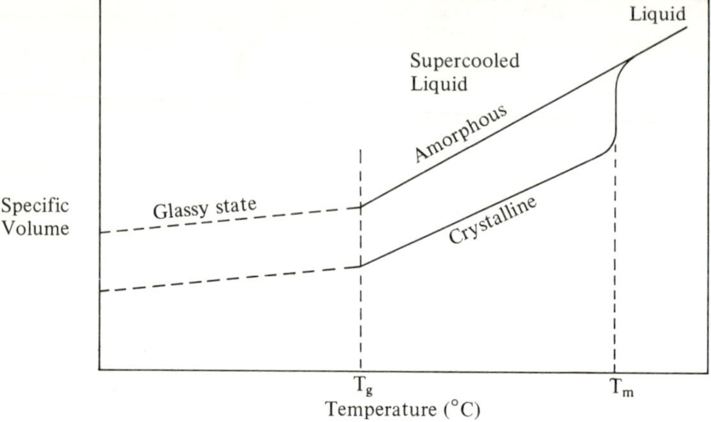

Fig. 26. Specific volume as a function of temperature for both amorphous and crystalline polymers

Let us consider the nature of these crystalline polymers (e.g. polyethylene) and examine how the polymer chains are arranged in the solid state. The earliest model of a crystalline polymer is shown in Fig. 27a in which different chains are arranged so that areas of regular packing are produced (known as the *fringe micelle structure*).

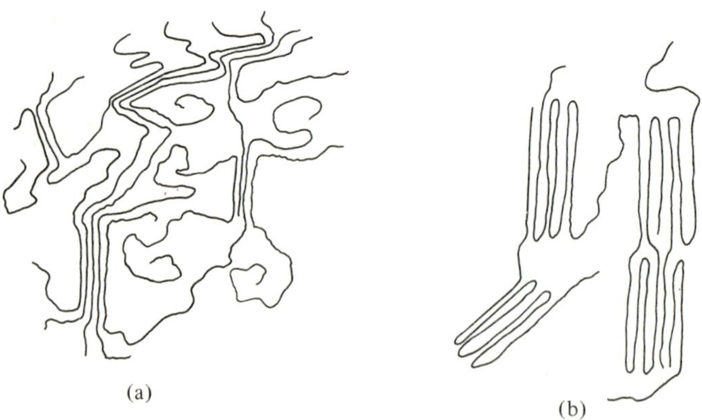

Fig. 27. (a) Fringe micelle model and (b) folded chain model of polymer crystallization

Later models favour the *folded-chain structure* in which areas of crystallinity are produced by one polymer chain folding backwards and forwards on itself to produce the regular pattern, Fig. 27b. These regions of atomic order are enmeshed in a matrix of randomly distributed chains so that the degree of crystallinity depends upon the relative amounts of these two components.

Crystallization is only one of the two principles that have been used to strengthen polymers. The other principle is *chemical cross-linking,* in which an element such as sulphur is added to the polymer to bond the different chains together by a chemical reaction rather than by physical orientation as in crystallization. A good example of a cross-linked polymer is rubber which is used for motor car tyres, in which the hardness is a function of the degree of cross-linking; if the cross-links introduced are numerous, then the very hard form of rubber known as *ebonite,* which has a very high softening temperature, is produced. If a polymer forms a very large number of cross-links on cooling the material will set as a rigid three-dimensional structure which will not soften upon reheating; such polymers are called thermosetting resins. Typical examples are bakelite (one of the original polymeric materials) and epoxy resins (such as araldite) in which the cross-links are formed by the addition of the hardener and moderate heating. Their structure and mechanical properties resemble the glassy polymers at very low temperatures, that is they have high strengths and moduli but are normally brittle. There is now a *third method* of producing a strong polymer, *chain stiffening,* in which the segments of the chain are made rigid. These segments are normally linked by rotatable bonds which can easily bend or fold on one another (like a jack-knifed lorry and trailer). If we stop this occurring by hanging very heavy groups of atoms on the chain, as in polystyrene, where benzene rings are attached to the carbon backbone, the softening temperature may be raised to 90°C without any crystallinity or cross-linking being introduced. The absence of crystallinity makes the material transparent, whilst absence of cross-linking makes it readily formable.

Obviously by combining these various mechanisms there are numerous possibilities of creating polymeric materials with quite different mechanical properties. Fig. 28 shows the three basic mechanisms at the corners of a triangle and how by *the combination of two or more* of these mechanisms, a wide range of polymeric materials can be produced.

If we are to understand the mechanical behaviour of polymers, we must also consider the temperature dependence of the elastic modulus, *for unlike metals the modulus is not a constant.* Fig. 29 illustrates this variation for a typical amorphous polymer; in the *glassy region* the modulus is relatively high (5000 MN m^{-2}) and the polymer is hard and brittle, the application of a stress producing little or no elastic strain.

Above T_g the material behaves like rubber and the modulus drops several orders of magnitude. This *viscoelastic region* is dominated by a leathery behaviour at low temperatures and rubbery behaviour at high temperatures (below T_f) which is due

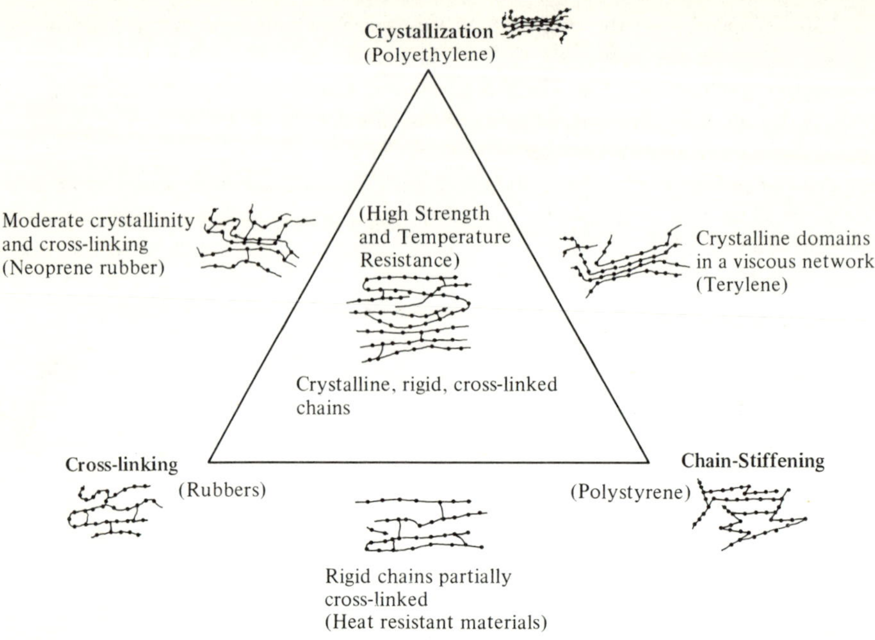

Fig. 28. Combined features make various properties of polymers possible. Each corner of the triangle represents a basic principle for making a polymer strong and resistant to temperature. The sides of the triangle and the centre indicate various combinations of these

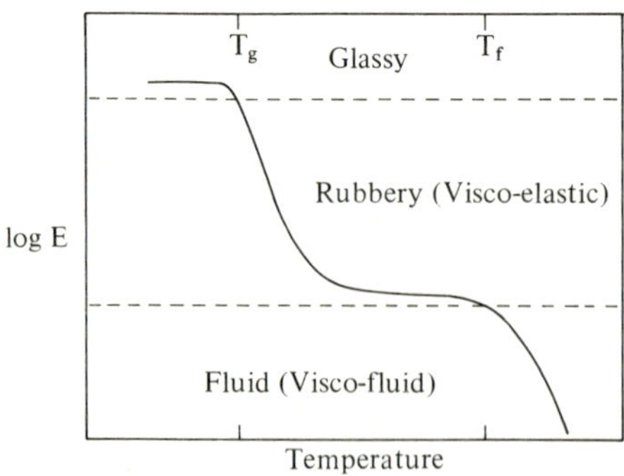

Fig. 29. The variation of Young's modulus with temperature for an amorphous polymer

to sliding and rotation occurring between the individual segments of the polymer. With true elastomers the plateau on the modulus curve is very distinct and the value of E may actually rise with temperature. At temperatures higher than the *viscofluid transition* (T_f) the modulus drops very rapidly and the application of a stress produces large irreversible strains in the material.

We shall now examine these effects in more detail as we consider the mechanical properties of specific polymer types.

Mechanical properties

As was stressed earlier in this chapter, the mechanical properties of polymers are very sensitive to the type of test, the temperature, and the speed with which the test is performed. Bearing these aspects in mind we can consider two tests that are used to measure the strength and ductility of organic polymers.

THE THERMOMECHANICAL TEST This method consists of determining the temperature dependence of the deformation caused by a constant stress under the same loading conditions. The normal practice is to measure the axial compression strain caused by a practically constant axial compression stress acting for a fixed period of 10 seconds. This measurement is carried out with a uniformly rising temperature (3°C/min) from – 150°C up to 300 to 400°C. The results are plotted as deformation-temperature curves as shown in Fig. 30.

The main features of these curves may be summarized as follows:

(1) Below the glass transition temperature (T_g) there is virtually no strain due to the complete locking of the polymer chains and their individual segments.

(2) Above T_g there is a rapid rise in the strain which is associated with a large reversible deformation of the polymer as the individual segments move relative to each other and the polymer chains uncoil.

(3) The third region (the viscofluid) is marked at T_f the flow temperature, with a very large increase in the strain which gives rise to large irreversible and reversible deformations as the polymer segments and chains move relative to each other.

Although in the viscofluid region *the polymer is useless as a structural material,* it is now in a form suitable for *drawing and the production of fibres.* Reverting to a consideration of the types of thermomechanical curves, Fig. 30b illustrates the thermomechanical curves for a typical elastomer (silicone rubber) and a typical hard plastic (polymethylmethacrylate – Perspex). An examination of these curves reveals that the region of the rubbery state of the elastomer covers the low and the slightly elevated temperatures, whereas the curve for perspex shows that it is still a glassy material at these temperatures. In the case of elastomers the T_{ge} determines

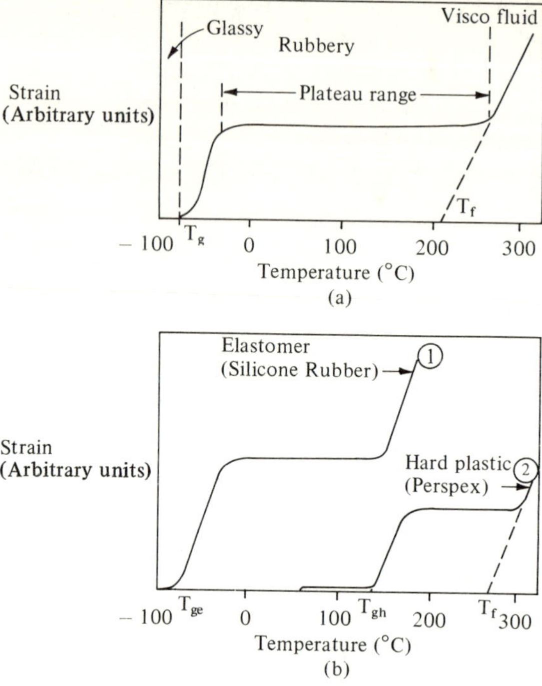

Fig. 30. (a) Thermomechanical curve for a typical amorphous polymer. (b) Thermomechanical curves for an elastomer (silicone rubber) and a hard plastic (polymethylmethacrylate). The glass transitions are shown at T_{ge} and T_{gh} respectively

the temperature region below which the polymer cannot exhibit elastomeric properties, whereas in the hard polymer T_{gh} determines the temperature above which the polymer softens and is a criterion of its heat resistance.

Experiment 12. Using the thermomechanical test obtain the curves for a typical elastomer and a hard plastic such as perspex. The test could be performed using a beam 80 to 100 mm long, 10 mm wide and 3 to 5 mm thick, which is supported on knife edges at the ends and the load applied to the mid-point of the beam. The deflection of the beam can then be measured with a dial gauge fitted with a long shaft to allow the beam to be immersed in alcohol cooled with dry ice and warmed with a hot air blower. Above room temperature use water, heated by a small immersion heater (such as those available for heating drinks).

Turning now to the crystalline polymers such as polyethylene or nylon, we can again consider the type of thermomechanical curves obtained. Figure 31 illustrates some of the typical results: curve 1 represents a highly crystalline polymer, so virtually no strain occurs below T_g and only a small amount above T_g, the large irreversible strains developing only at T_m the melting point of the crystallites. As this polymer has a melting point which is greater than the viscofluid temperature ($T_{f'}$), the thermomechanical curve continues to rise at T_m and above. If the viscofluid temperature $T_{f''} > T_m$ the polymer (curve 2) will enter a viscoelastic state above T_m and exhibit a plateau characteristic of the rubbery state between T_m and $T_{f''}$. As we noticed in Chapter 2, if a molten crystalline polymer is quenched it will become amorphous, whereas on slow cooling the polymer will exhibit crystallinity. If a thermomechanical test is carried out on a quenched polymer (polystyrene or polythene), on warming through the glass transition range it *enters the rubbery state* (curve 3). For most polymers the duration of the test is *too short* for much recrystallization to occur and the polymer will behave in a similar manner to an amorphous type going through a viscofluid transition at $T_{f'}$.

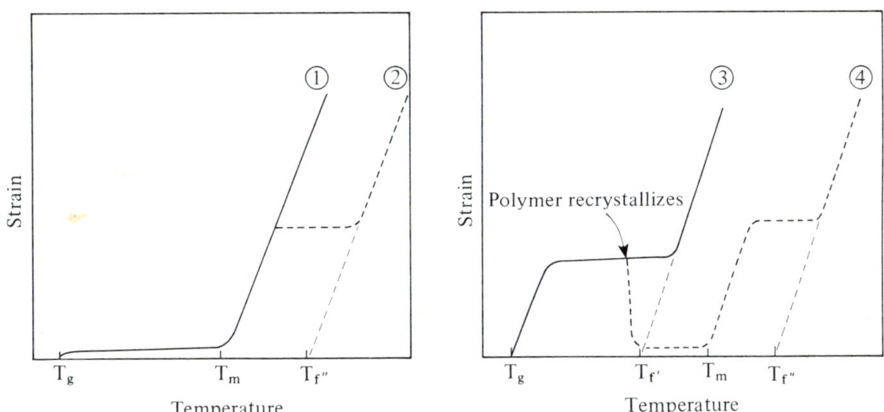

Fig. 31. Thermomechanical curves for crystalline polymers

If however recrystallization did occur before $T_{f'}$ the ability of the material to deform viscoelastically would be reduced and the *strain would drop* to a very small value (curve 4), it would then behave in a similar manner to the polymer whose viscofluid temperature is greater than the melting temperature of the crystallites (2).

Experiment 13. Carry out thermomechanical tests on a range of crystalline polymers, including high and low crystallinity nylon, polythene and mylar.* By carefully heating polythene on an electrical hot-plate it should be possible to alter the degree of crystallinity of the material by either quenching or slowly cooling it. If the quenched material is annealed at an intermediate temperature it will be possible to test it in a recrystallized form.

* Some of the materials available from G. H. Blore Ltd., 480 Honeypot Lane, Stanmore, Middlesex.

The changes in state detected by the thermomechanical test can be correlated with the changes in the elastic modulus. For the highly crystalline form shown in Fig. 32, there is only a small change on passing through the glass transition, this being followed by a gradual decrease in modulus until it reaches T_m at which it drops rapidly with the melting of the crystallites.

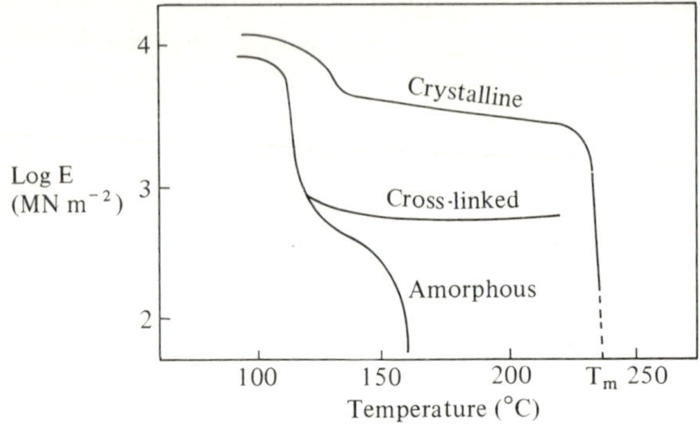

Fig. 32. The variation of Young's modulus with temperature for crystalline, cross-linked and quenched (amorphous) polystyrene

For the cross-linked polymer there is a drop in modulus of over three orders of magnitude during the glass transition, but then it remains *relatively constant* during the viscoelastic region only decreasing rapidly when the viscofluid transition is reached. The quenched form (amorphous) behaves exactly as the true amorphous polymer (see Fig. 30). Thus both *cross-linking* and *crystallization* may be used *to raise the softening temperature* and hence the working temperature of polystyrene.

STRESS-STRAIN CURVES We saw in Chapter 3 how the simple tensile test could be interpreted in terms of the structure of the metallic material; when we come to consider polymers, similar structural models can be used to explain the shape of the curves obtained if it is realized that the exact shape is *extremely sensitive to temperature and strain rate.* Fig. 33a illustrates some of the curves obtained for amorphous polymers. Well below the glass transition the material is completely brittle (curve 1) and fracture occurs on the elastic loading line. Although it would appear to be possible to calculate E from these curves, it must be remembered that the *strain is time-dependent* at all but the lowest temperatures, hence the values calculated for E will be high. At temperatures close to and below T_g the material undergoes '*forced rubbery elasticity*'.

After deforming elastically (curves 2 and 3) to the yield stress σ_y, the stress drops to a lower value σ_d (the draw stress) and *a neck appears* in the specimen. Further deformation causes the neck to propagate along the specimen and the stress only increases again when the whole specimen has been drawn down by the neck. The formation of the neck (which is the strongest part of this material, in contrast to being the weakest part of a metal) is probably due to the applied strain energy causing a lowering of the modulus by raising the temperature at that point. This will allow the polymer chains to orientate themselves in the direction of the applied stress

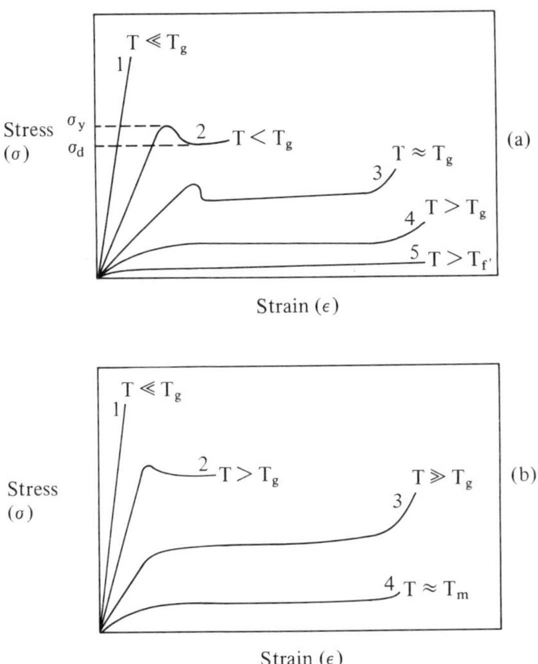

Fig. 33. The stress-strain curves for (a) amorphous polymers and (b) crystalline polymers at different temperatures (temperature increases with ascending numbers of curves)

and hence become much stronger. If the material remains at or below T_g these oriented chains remain frozen in even though the stress is removed. They will however revert to their random distribution if the material is gently warmed above T_g. At temperatures above T_g (curve 4) the large reversible elastic strains can develop from the beginning of the test and no yield point is observed. If the temperature of the test is greater than T_f, viscofluid behaviour occurs with chain sliding dominating the large irreversible strains developed (curve 5).

In the case of the crystalline polymers (Fig. 33b) we observe a similar set of curves, although the *structural explanation is rather different.* At temperatures well below T_g (curve 1) the material is similar to the amorphous polymer and a pseudo-Hookean behaviour is observed. Whereas an amorphous polymer can exhibit forced rubbery elasticity below T_g, the crystalline polymer is still brittle up to this temperature. Above T_g (curve 2) the curve is similar to No. 2 of the amorphous polymer and a neck will be observed in the specimen; however *the mechanism of neck formation is quite different.* In a crystalline polymer the neck is produced by recrystallization of the chains in the direction of the stress owing to the fact that the melting point of the aligned chains is higher than the unaligned chains. These unaligned chains will therefore melt and the neck will progress along the sample, but unlike the amorphous polymer no rearrangement will occur if the material is gently warmed above T_g. The propagation of the neck is usually terminated at a temperature around T_g by flaws in the actual sample. At higher temperatures (curve 3) the neck propagates along the entire length of the sample and the stress rises again when the aligned polymer chains become strained. At temperatures close to the melting point of the crystallite T_m (curve 4) the material behaves in a similar viscofluid manner to the amorphous polymer.

Experiment 14. Using the tensile machine and temperature enclosure discussed in Chapter 3, obtain the stress-strain curves for a wide range of amorphous and crystalline polymers. In particular, examine the effect of annealing the samples after they have been deformed in the viscoelastic state (i.e. below T_g for amorphous polymers and above T_g for crystalline polymers). Using similar material to that produced for the thermomechanical tests, examine the effect of the degree of crystallinity. It is suggested that the specimens be made from sheet material which is readily available and easy to fabricate. As polymeric specimens are not very stiff, it may be desirable to provide some supports inside the temperature enclosure to take the weight of the chucks.

Time dependence – creep and stress relaxation

We have had to remind ourselves frequently that *time is a very important parameter in the testing of polymeric materials*; we can now consider this variable in a little more detail as we discuss the *creep* and *stress relaxation* of polymers in their viscoelastic state.

Creep is defined as the continuing extension of the material with time when a constant load is applied. The effect is quite small for metals at normal temperatures and only becomes serious with high temperatures; for polymeric materials, creep is important even at room temperature owing to their inherently low melting points. Various mechanical models have been suggested to describe this behaviour, the simplest consisting of an elastic spring and a dashpot (this is a cylinder with a close-fitting piston which is filled with a fairly viscous liquid that can flow past the piston as it is moved: a syrup tin fitted with a plunger is a good example of this).

If a load was applied to this system obviously the dashpot would deform rapidly and transmit the strain to the spring. Therefore we would expect an initial rapid straining of the specimen followed by a period in which the strain rate becomes less. A normalized plot of the variation of strain with time as illustrated in Fig. 34a confirms that this is what occurs. Obviously temperature, as well as load, is a controlling factor and enters into our model by altering the value of the viscosity of the liquid in the dashpot (heated syrup is much runnier than cooled syrup). *Creep is probably the most serious limitation to the use of polymeric materials at temperatures much in excess of 100 to 150°C.*

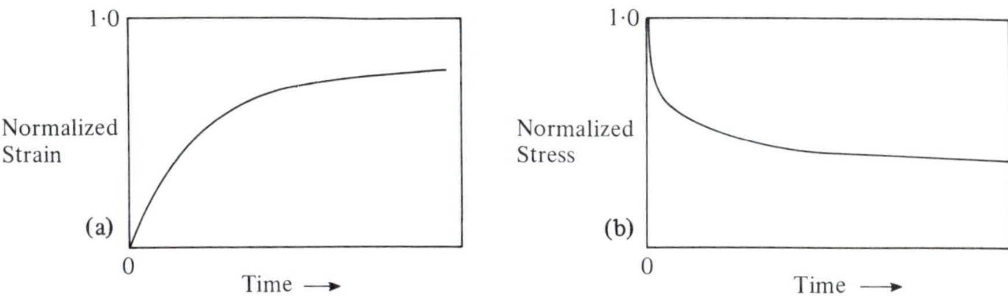

Fig. 34. (a) The strain-time relationship for idealized creep and (b) idealized stress relaxation.

Stress relaxation, by which a material strained to a given deformation can reduce the applied load, may also be described by a similar model; for this model there is an exponential decay in the stress (σ) with time (t) at a constant value of strain which is governed by the stress relaxation equation:

$$\sigma = \sigma_0 e^{-t/\theta} \tag{4.1}$$

where θ is a constant and σ_0 is the initial stress. Fig. 34b illustrates the normalized stress as a function of time. Stress relaxation is a serious problem in the use of polymeric materials for structural applications, for example nylon nuts and bolts tightened at low temperatures when subjected to slightly higher temperatures will slowly relax the applied torque and the assembly will become loose.

These two models are *gross simplifications* of the real time dependence of strain and stress in polymeric materials and complex models are required to describe their behaviour adequately.

Summarizing this chapter on polymers, the important thing to remember is the extreme sensitivity to temperature, structure, and strain rate of the simple mechanical properties such as tensile strength, elongation, and reduction in area; but it is precisely this critical dependence of properties which has allowed the wide range of polymeric materials from very soft rubber to hard brittle glasses to be developed.

Experiment 15. Measure the creep behaviour of a polythene rod specimen in the temperature range 20 to 60°C by applying a fixed load and measuring the strain developed with either a strain gauge or the optical displacement of markers on the surface. By examining a range of applied loads and temperatures, estimate which has the larger effect on the creep properties.

Experiment 16. The stress relaxation behaviour of polythene can also be measured in the temperature range 20 to 60°C. The major problem here is to devise a suitable way of recording the drop in stress from the initial value as a function of time. (Hint: One way of achieving this would be to have the load-measuring device of the tensile machine used to plot out directly on a rotating drum which is started when the initial stress is applied. If it is impossible to devise an automatic recorder, the experiment can be performed if one person calls out the load reading and another the time reading into a tape recorder which can then be played back for the analysis of the results.) Compare your results with those predicted from equation 4.1.

5 Ceramics and glasses

Structure and constitution

Unlike the polymeric materials we discussed in the previous chapter, both ceramics and glasses have been used by man since the dawn of history. Man soon appreciated the *hardness,* the *low thermal conductivity* and *high corrosion resistance* of the stones that surrounded his cave. He soon learned that, although they were hard, they were easily split and that it was possible to fashion them into simple tools and weapons. He also quickly discovered that the clay which he walked on was easily shaped and that heating it in the fire turned it into a hard but porous pot, and further heating with the addition of a low-melting-point glass allowed him not only to produce a non-porous article but an artistic and aesthetically pleasing one. In Chapter 2 we saw that the structure of a ceramic was based upon the *ionic bonding* of elements such as silicon, magnesium, and aluminium with oxygen (other elements besides oxygen can form ceramics, such as nitrogen, boron, and carbon). In the case of silicon and oxygen we saw that they bonded together to form a tetrahedral unit (Fig. 11) which could be used as a subsequent building block. The organization of these tetrahedra will control the properties of the final product, if they are built up in a *three-dimensional array* the strong material quartz is produced; if, however, we joined them end to end a *chain* structure would result which is the basis of fibrous materials such as asbestos (Fig. 35a). The tetrahedra can also be arranged to form *sheet* structures such as those found in minerals like talc and mica in which the sheets can be easily split off or slid over each other.

Clay is also based on the silica tetrahedra, but in this case the silica is combined with aluminium oxide to form the mineral *kaolinite* ($Al_2O_3 \cdot 2SiO_2 \cdot 2H_2O$). Kaolinite is made up of irregular hexagonal plates about 5000 Å wide and 300 Å thick, each plate consisting of alternate layers of silica and alumina held together by the hydroxide ions (Fig. 35b). When the mineral is wet the sheets slide easily over each other and give rise to the observed plasticity of clay, but if the water is driven off the platelets become interlocked and the clay becomes rigid, as occurs in a hot summer. If the clay is fired the remaining water is removed and the silica combines with impurities to form a liquid glass that glues the platelets together and produces a relatively strong material (compressive strength 687 MN m^{-2}).

We have noticed in the case of metals that defects in the crystal packing caused them to be much weaker than they should be in theory. The same state of affairs exists in ceramics, the defects being vacancies (i.e. empty lattice sites), crystal boundaries which distort and rupture the semidirectional ionic bonds, and mismatching of ions that allows ions with the same charge to be brought together and hence to create areas of local repulsion. In very pure single crystals of ceramics, such as lithium fluoride, dislocations similar to those in metals have also been found, but they are unlikely to have a significant effect on the properties of normal ceramics.

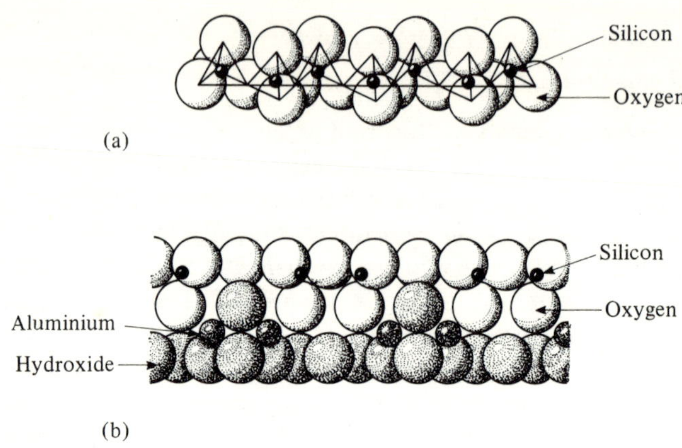

Fig. 35. (a) Silicate chain structure and (b) kaolinite sheet structure seen from the side as a 'laminate' of silicate and aluminate sheets (shaded)

A large number of *non-oxide ceramics* have become available recently to meet the needs of space technology and high temperatures. The carbides are particularly attractive due to their inherently high melting points (see Fig. 1). Silicon carbide and tungsten carbide have been used extensively as abrasives and cutting materials (high-speed lathe tool-tips). Another new ceramic is silicon nitride (tensile strength 137 MN m^{-2}) which has been considered for rocket nozzles and high-temperature gas flow pipes. Because of their very high melting points and hardness these types of ceramics have to be fabricated by sintering under pressure and the strength will depend on the grain size, degree of purity, and the degree of sintering.

When we were discussing polymers we noticed that at low temperatures they were converted into a *glassy state,* the essential feature of which was that the polymer chains were rigidly arranged in a random distribution. A similar state of affairs exists in inorganic glasses based on silica and alumina and other oxides; and, as we saw in Chapter 2, the atomic arrangement in a glass is one in which only a certain amount of short-range order exists (Fig. 11b) and its structure is essentially the same as a *supercooled liquid.* The original liquid becomes particularly viscous near to the freezing point and the formation and growth of nuclei is prevented. As the viscosity increases further with falling temperature the molecular arrangement lags further and further behind the true equilibrium arrangement and obviously the mechanical properties will therefore strongly depend upon the thermal history as well as the composition of the glass.

Mechanical Properties

ROOM TEMPERATURE For most ceramics and glasses we can say that at room temperature they are *hard and brittle* and exhibit elastic behaviour up to the point of fracture (the exceptions to this are the pure single crystals containing a few dislocations which have an elastic-plastic yield similar to metals), but the observed fracture strength (150 to 350 MN m^{-2}) of these materials falls short of the theoretical strength (> 20 000 MN m^{-2}) as is shown in Table 3.

Table 3 Strength of selected ceramics, glasses, and clays

		Strength (MN m^{-2})	
Material	Structure	Tension	Compression
Magnesium oxide	Polycrystalline (grain size 30 μm)	194	–
Aluminium oxide	Polycrystalline ceramics	206	3100
Beryllium oxide	Polycrystalline (grain size 10 μm)	284	–
Magnesium oxide	Single crystal (defect free)	1100	–
Silicon carbide	Polycrystalline	171	1050
Tungsten carbide	Polycrystalline	343	540
Silicon nitride	Polycrystalline	137	600
Clay	Fired (1300°C) crystalline composite	–	79
Porcelain	Fired (1600°C) semi-glass crystalline	–	687
Soda glass	Amorphous	~ 70	–
Borosilicate glass	Fibres (10 μm diameter)	3730	–
Silica	Fibres (25 μm diameter)	13 800	–
Carbon	5 μm whiskers single crystal	19 600	–

We saw in the case of the yielding of metals that the *discrepancy* in the experimental and theoretical values was due to the presence of the dislocation; in the case of ceramics this discrepancy is due to **surface flaws and cracks** (known as Griffith cracks). The effect of these cracks was clearly demonstrated in an experiment on a MgO polycrystalline ceramic; when the surface of the material had been chemically polished to remove the surface defects, the tensile strength was 194 MN m^{-2} (for a 5 μm grain size), this value is of course much lower than the tensile strength of a chemically polished single crystal (1100 MN m^{-2}) owing to the fact that the grain boundaries in the polycrystalline material act as surface flaws themselves. If the polished crystalline material was sprinkled with silicon carbide powder the fracture strength decreased to 129 MN m^{-2} owing to the flaws introduced.

The theoretical fracture stress (σ^*) for a totally elastic body may be calculated using a similar model to the one we used to calculate the theoretical yield strength of a metal. This gives the value of σ^* as:

$$\sigma^* = \left(\frac{E\gamma}{a}\right)^{\frac{1}{2}} \tag{5.1}$$

where E is the elastic modulus, γ the surface energy of the crack formed, and a the original interplanar distance.

The theoretical value predicted by this equation (20 000 MN m^{-2}) will only be achieved if there are *no cracks present* in the material prior to fracture. If there are cracks (Fig. 36), then the tip of the crack will act as a stress concentration and the

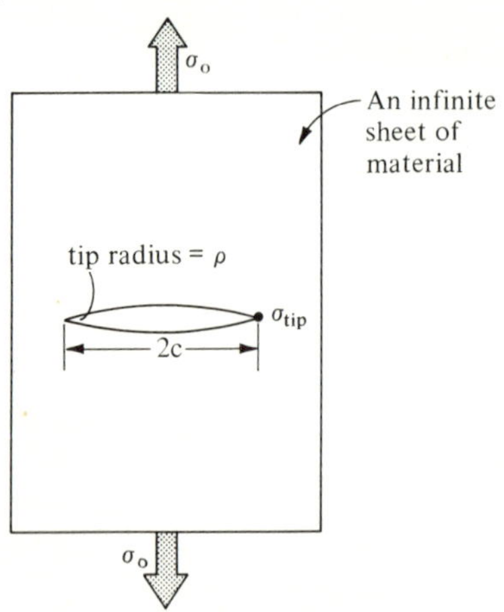

Fig. 36. Stress concentration σ_{tip} at a crack length 2c in an elastic solid stressed at σ_o

tip stress will equal the theoretical fracture strength well before the bulk stress reaches this value. The tip stress σ_t is related to the applied stress σ_o by $\sigma_t = 2\sigma_o(c/\rho)^{\frac{1}{2}}$ where $2c$ is the crack length and ρ is the tip radius. With a tip radius

of 4 Å the stress concentration factor (σ_t/σ_o) would exceed 100 and thus cause the failure of the material.

If we assume that the tip stress equals the theoretical fracture stress at failure we may relate the crack size to the measured fracture stress σ_f as follows:

$$\sigma_f = \left(\frac{2\gamma E}{\pi c}\right)^{\frac{1}{2}} = Kc^{-\frac{1}{2}} \tag{5.2}$$

Although we have been considering ceramics, this equation applies equally to glass and glassy polymers.

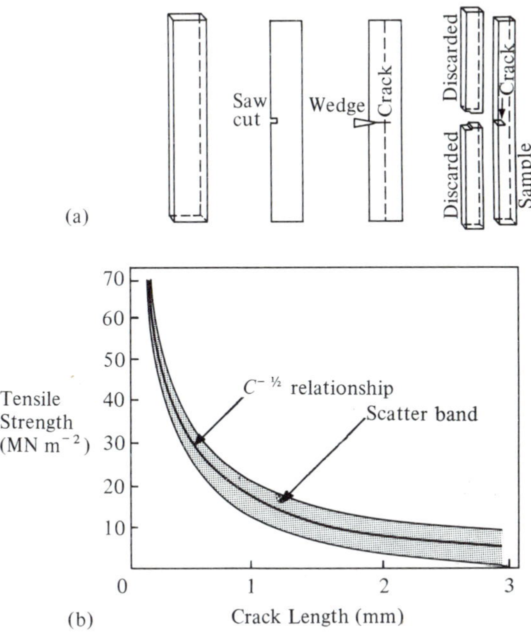

Fig. 37. (a) Method of preparing tensile specimens and (b) the dependence of tensile strength on crack length in perspex ($E\gamma = 5{\cdot}0 \times 10^{11}$ N^2 m^{-3})

We may therefore conclude that the mechanical behaviour of ceramics at normal temperatures is *controlled by the presence and nature of these surface flaws.*

Experiment 17. Using a glassy polymer (perspex is ideal) the relationship between fracture strength and flaw size may be investigated. Figure 37a indicates how specimens containing initial cracks can be obtained. If these are tested in simple tension the fracture strength can be measured and if the equation (5.2) is valid the results should follow the curve indicated in Fig. 37b for a set of perspex specimens.

Glass provides an interesting material for the study of these surface effects, since the mechanical properties of glass are controlled to a large extent by the manner in which it is corroded by water. Most glass failures occur because of a combination of stress and corrosion known as *static fatigue.* The mechanism of this failure involves the migration of the sodium ion, which leaves a free oxygen bond to combine with a water molecule forming silica hydroxide and an hydroxide ion:

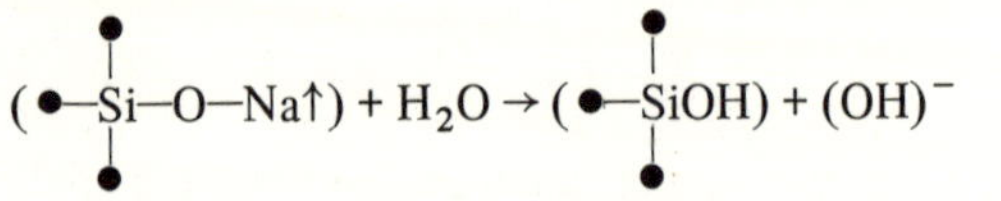

$$(\bullet\!-\!\underset{\bullet}{\overset{\bullet}{\mathrm{Si}}}\!-\!\mathrm{O}\!-\!\mathrm{Na}\uparrow) + \mathrm{H_2O} \rightarrow (\bullet\!-\!\underset{\bullet}{\overset{\bullet}{\mathrm{Si}}}\mathrm{OH}) + (\mathrm{OH})^-$$

The free hydroxide ion further combines with silica to produce:

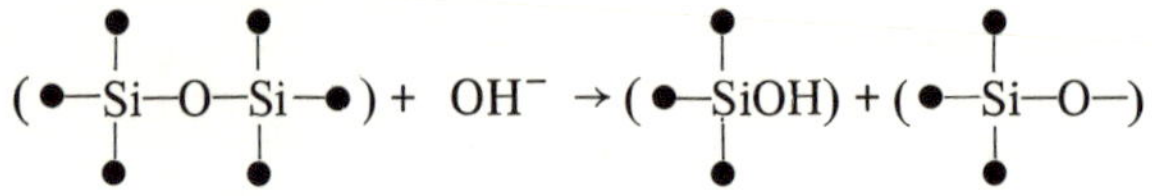

$$(\bullet\!-\!\underset{\bullet}{\overset{\bullet}{\mathrm{Si}}}\!-\!\mathrm{O}\!-\!\underset{\bullet}{\overset{\bullet}{\mathrm{Si}}}\!-\!\bullet) + \mathrm{OH}^- \rightarrow (\bullet\!-\!\underset{\bullet}{\overset{\bullet}{\mathrm{Si}}}\mathrm{OH}) + (\bullet\!-\!\underset{\bullet}{\overset{\bullet}{\mathrm{Si}}}\!-\!\mathrm{O}\!-)$$

The reaction then continues:

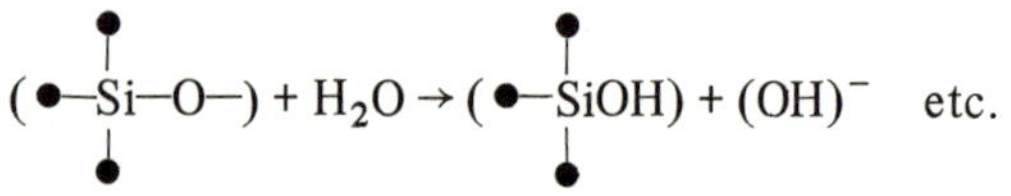

$$(\bullet\!-\!\underset{\bullet}{\overset{\bullet}{\mathrm{Si}}}\!-\!\mathrm{O}\!-) + \mathrm{H_2O} \rightarrow (\bullet\!-\!\underset{\bullet}{\overset{\bullet}{\mathrm{Si}}}\mathrm{OH}) + (\mathrm{OH})^- \quad \text{etc.}$$

These reactions increase the pH of the corrosion layer making the dissolution autocatalytic. Slowing down the rate of pH increase leads to a decreased dissolution, explaining why water corrosion is less damaging than steam vapour, and why the addition of an acid will cause less dissolution.

Experiment 18. An interesting experiment to illustrate that the sodium ions can become mobile and conduct electricity in glass is to heat a glass rod in a fish-tail bunsen flame. Connect wires to the ends of the rod and couple one direct and the other through a low-wattage bulb to the mains supply through an isolation transformer. As the current begins to flow through the glass rod the bulb will glow.

Experiment 19. If a soda glass rod (4 mm diameter) is clamped firmly at one end using aluminium foil to protect the clamped area, the effects of different surface treatments can be studied by gradually loading the rod at the end and noting the bending load for fracture. Compare surfaces which have been scratched, chemically etched with HF, and flame polished (draw the rod through a bunsen flame so that it only just begins to soften). Use a source of heated steam to investigate the static fatigue of the rods in the flame-polished and lightly scratched conditions.

If we treat the surface of the glass so that a *compressive stress is developed,* it means that the cracks in the surface have to overcome the extra compressive stress before brittle failure can occur. The *toughening of glass* can either be accomplished by cooling the glass rapidly in a cold air stream from above its softening temperature or heating the glass in a molten salt bath (potassium nitrate/nitrite eutectic), where an ion exchange mechanism leads to the replacement of the sodium ions by larger potassium ions and a consequent build-up in the surface compressive stress.

Experiment 20. Repeat the previous experiment using a glass rod which has been toughened by either the chemical or thermal treatments and compare the increase in the strength of the glass. What is the relevance of this experiment to the manufacturer of car windscreens?

ELEVATED TEMPERATURES At temperatures near to the melting point of the ceramic, the material becomes weaker as illustrated in Fig. 38 and can deform by a creep mechanism, similar to that observed in metallic materials.

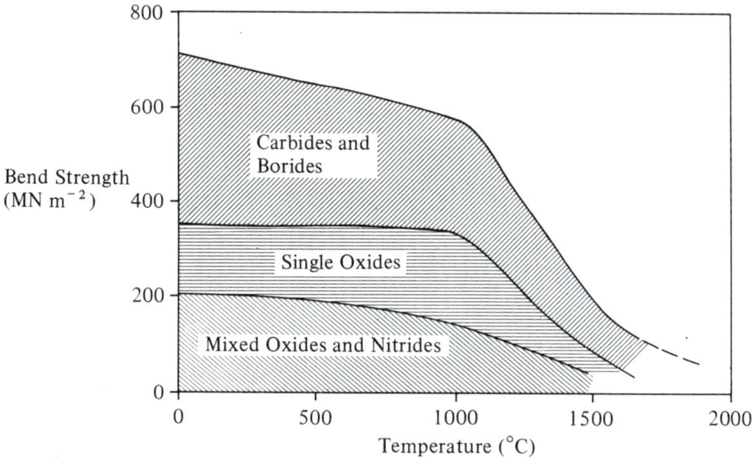

Fig. 38. Bend strength as a function of temperature for various ceramic materials

The ionic bond becomes less stable at high temperature and the increased thermal vibrations will allow the ions to migrate under the influence of stress or an applied electric field. The general equation governing the strain rate is as follows:

$$\dot{\epsilon} = A\sigma^{n} e^{-Q/kT} \tag{5.3}$$

where $\dot{\epsilon}$ is the strain rate, σ the applied stress, and Q the activation energy for the process. (For viscous flow or grain boundary sliding the value of the exponent would be 1.) The creep rate is not only dependent upon the stress level and temperature but also upon the grain size of the ceramic, fine-grained ceramics appearing to have a higher creep rate owing probably to the larger grain boundary area.

In comparison with metals, a ceramic of equivalent tensile strength to a metallic material will have superior creep properties, a fact that is utilized in the production of nimonic alloys in which ceramic particles are deliberately introduced into the metallic matrix to hinder dislocation movement. In glassy materials the lack of long-range order makes any simple deformation process, even at elevated temperatures, impossible. However these materials can deform by the *viscofluid process* we discussed in relation to the behaviour of polymers above their glassy transition. Figure 39 illustrates the variation of the viscosity of a soda glass with temperature.

Notice that below T_g the viscosity deviates from the predicted curve because the behaviour is no longer truly viscous, but is now viscoelastic. In the viscous state the viscosity η is given by:

$$\eta = TA e^{-Q/kT} \tag{5.4}$$

whereas in the viscoelastic state we must use a model similar to that proposed for the viscoelastic behaviour of polymers.

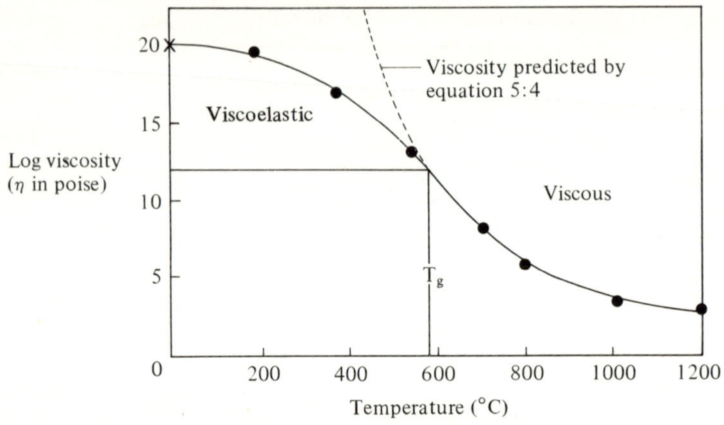

Fig. 39. The variation with temperature of the viscosity of a soda glass (the glass transition temperature is 590°C)

In the melting range 800°C and above, the viscosity is very low and allows glass to deform rapidly under its weight. The intermediate values of the viscosity obtained at temperatures of 450 to 650°C allow the glass to be formed by procedures such as blowing, rolling, or drawing.

Experiment 21. The mechanical behaviour of soda glass may be investigated if a long rod is clamped firmly at one end and a fixed weight hung on the other. If the rod is heated by a small resistance furnace situated in the middle of the rod, the strain (ϵ) produced in a given time (t) by the weight plus the weight of the rod below the heated section can be measured. A simple way of measuring strain is to fix a dial gauge so that it records the deflection of the weight at various temperatures. For an ideal Newtonian fluid the shear stress $\tau = \eta \partial\epsilon/\partial t$, thus the viscosity can be measured approximately at each temperature.

The glasses we have considered so far have been amorphous, but there are now *glass ceramics* in which crystallization has occurred by a nucleation and growth reaction. Some ordinary glasses will partially crystallize after prolonged exposure to sunshine (*devitrification*) but in these glass ceramics the crystallization is induced by adding catalysts such as colloidal gold, silver, and copper which act as nuclei to produce a totally crystalline structure. Various aluminosilicate glasses are suitable especially those containing Li_2O. Because of the fineness of the colloidal phase the final crystal size is very small ~ 0·01 μm which is several orders of magnitude less than sintered ceramics. These glass ceramics have *attractive mechanical properties* which approach those of normal ceramics and a very low thermal coefficient of expansion which makes them eminently suitable for very large glass objects such as telescope mirrors.

Modern man therefore still uses these ceramic materials: he builds large dams with clay cores, such as the Volta dam in Ghana. Thousands of tons of concrete, which essentially consists of ceramic particles mixed with a glue (e.g. Portland cement which is a complex calcium silicate), are used every year in numerous building projects and, as he pushes out into space and develops new power units, he utilizes the superior high-temperature properties of these crystalline ceramics.

Appendix: Mechanical testing

Various types of mechanical tests, some of which are illustrated in Fig. 40, have been devised in order to estimate the mechanical properties of different materials. Some of these tests can be used on all materials such as the tensile and hardness tests, whereas other such as the compression test are only normally used for brittle materials such as cast iron and ceramics.

Let us now consider some of these tests in a little more detail.

The tensile test

The tensile test, for which a typical stress-strain curve is shown in Fig. 41, is the most widely used test in industrial and research organisations. For reasons of convenience we usually choose to plot the approximate stress-strain curve (that is, the load/original area of the sample measured in Nm^{-2} and the change in length/original length x 100% of the sample).

Another convenient way of plotting the experimental data is to record the load versus extension of the sample; the extension being measured either by the separation of the machine crosshead or by an extensometer clamped to the specimen under test. Comparison of the approximate and true stress-strain curves in Fig. 41 shows that the two methods of plotting the data obtained from the tensile test give the same result for the initial portion of the curve, but *deviate rapidly* from each other *beyond the maximum load.*

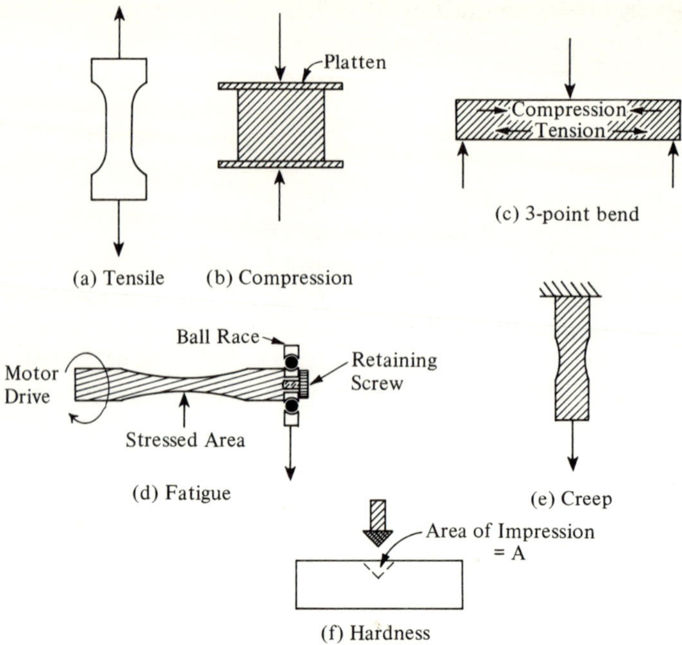

Fig. 40. Various mechanical testing methods

In engineering practice the maximum load/original area is quoted as the *ultimate tensile strength* and is the value usually given in this monograph. The true stress continues to rise after the maximum load has been reached owing to the reduction in the cross-sectional area of the specimen in the necked region by the large amount of plastic deformation occurring.

In addition to information on strength, the tensile test supplies data on the *ductility* of the material. The percentage elongation is the most widely used measure of ductility and is defined as (the increase in gauge length/gauge length) × 100% (where the gauge length is measured on the untested and reassembled bar after fracture). The normal practice is to have the gauge length = 4 × the diameter of the specimen. The *percentage elongation* obtained from these measurements is a sum of both the uniform and nonuniform (necking) deformation. However with long gauge lengths the nonuniform deformation is relatively small.

With a ductile material which has a relatively large plastic deformation it is often difficult to define from the load-extension curves the elastic limit (or limit of proportionality) with any certainty, and thus other methods of defining this value are adopted. The simplest is to draw a line parallel to the initial portion of the curve (see Fig. 41) so that it passes through a chosen offset such as 0·1% strain. Where this line intersects the load curve the load value/original area is called the *0·1%*

proof stress (other values of proof stress may be given such as 0·2 or even 1·0% if required).

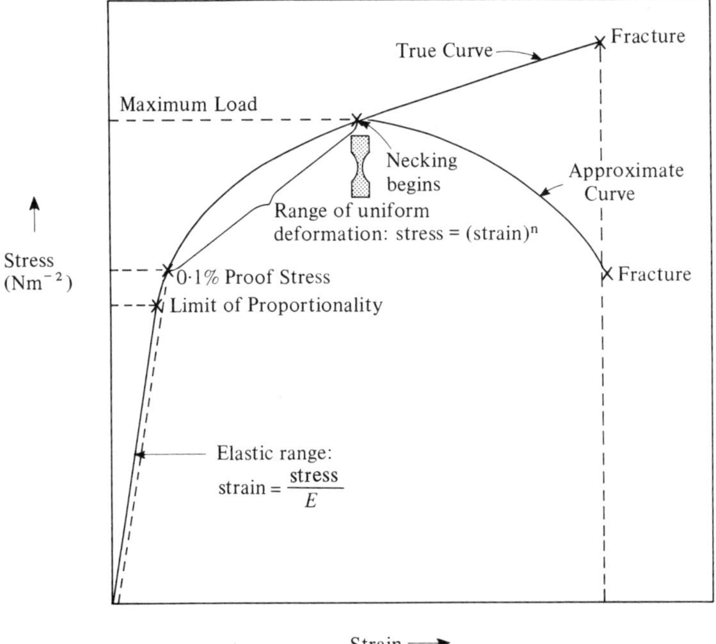

Fig. 41. The tensile stress-strain curve indicating the different ways of plotting the results, and the significant points on the curves for a ductile metal sample

The compression test

The uniaxial compression test can give information similar to the tensile test and is particularly valuable when dealing with highly brittle materials such as ceramics and glasses. The disadvantage of the compression test is that frictional forces are developed between the compressing platten and the specimen, which make the analysis of the stress within the specimen difficult.

Hardness testing

It is possible in many cases to substitute for the relatively slow and expensive tensile test a more convenient measure of the strength of a material, that is a hardness test. *Hardness* is defined as the resistance to penetration and the majority of hardness testers force a small sphere, pyramid, or cone into the surface of the material by means of an applied load. The hardness number is defined as the applied load/area of the impression formed, and quoted in kg mm^{-2}.

The *Vickers Hardness* number (VHN) which is based on a diamond pyramid indentor varies from 20 to 30 for plastics and pure metals to 2000+ for some ceramics. The *Brinell Hardness* number (BHN) is based on a ball indentor and is similar to the diamond test but becomes invalid above VHN = 350 owing to the deformation of the steel ball.

Bend tests

Bend tests are commonly used in engineering where beams and struts are concerned. Because the bent specimen is subjected to both tensile and compressive stresses, the interpretation of the results is rather difficult; shearing stresses are also present, so that a material weak in shear may fail prematurely and record a low tensile strength value. These tests are valuable for brittle materials such as ceramics and an adapted form is used in the thermomechanical test for polymers.

Creep and fatigue

The measurements in the tests so far described are short time measurements, i.e. the strain is measured within a few seconds of applying the load. The *creep test,* however, records the strain as a function of time for a constant stress (or constant load which is simpler) over a range of temperatures. After the initial instantaneous strain on the application of the load, a further time-dependent strain is observed which has great practical significance in load-bearing and precision equipment.

Fatigue testing normally uses a periodically varying stress and the usual test employs a bar rotated about its longitudinal axis with a load applied through a bearing at one end. One element of the bar's surface is therefore subjected to an alternating compression and tension in each cycle. Other fatigue machines employ a push-pull arrangement which has the advantage of allowing a mean stress to be applied on top of the alternating stress – for example to simulate the fatigue behaviour of a car axle subjected to the weight of the car plus the vibrations from the road surface.

Temperature enclosures

It is often very valuable to extend the range of the tensile test to different temperatures in the range −150°C to 100°C. This can be done in the simple enclosure illustrated in Fig. 42a; if the chuck rods are a good sliding fit in the paxolin box very little liquid seepage will occur. It is important to ensure that the enclosure is carefully aligned on the tensometer so that it does not interfere with the axial loading of the measuring beam.

Figure 42b illustrates a similar type of box which can be used for the thermomechanical test on polymeric materials.

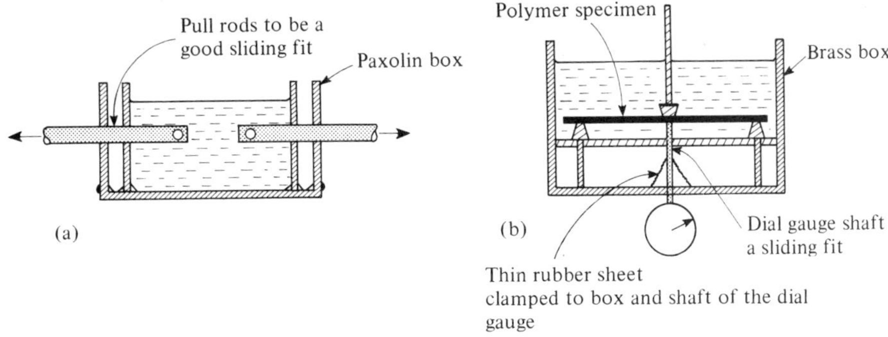

Fig. 42. Design of simple temperature enclosures for (a) tensile testing and (b) three-point bend testing

Bibliography

1. GILLAM, E. *Materials under Stress.* Newnes-Butterworth, 1969.
   A good book which deals with the complete range of materials at a fairly advanced level.
2. A Scientific American Book. *Materials.* W. H. Freeman, 1967.
   A series of very well-written articles covering all aspects of materials in a fairly general manner. Good background reading.
3. MOFFAT, W. G., PEARSALL, W., and WULFF, J. *The Structure and Properties of Materials, Vol. 1 Structure.* John Wiley, 1965.
   A book that deals with the structure of materials at 1st year University level.
4. ALEXANDER, W., and STREET, A. *Metals in the Service of Man.* Pelican, 1968.
   A general book that covers all aspects of metals, it has useful chapters on the extraction of metals and the properties of specific alloy systems.
5. BIGGS, W. D. *The Mechanical Behaviour of Engineering Materials.* Pergamon (Commonwealth and International Library), 1965.
   This has a good chapter dealing with testing methods.

6. KEMPSTER, M. H. A. *Teach Yourself Engineering Materials.* English University Press, 1968.
This book is primarily about metals and deals with them from a practical point of view. Good chapters on specific alloy systems.
7. PALIN, G. R. *Plastics for Engineers.* Pergamon (Commonwealth and International Library), 1967.
A very useful book which deals with both the chemistry and the properties of polymeric materials.
8. METALS AND MATERIALS. The Institute of Metals.
A journal which is published monthly by The Institute of Metals. It contains excellent articles describing various simple pieces of apparatus and experiments with materials. The author will be happy to supply a full set of references on request.

Index